SIGNS OF HOPE

SIGNS OF HOPE

Working Towards Our Common Future

LINDA STARKE

The Centre for Our Common Future

Foreword by

GRO HARLEM BRUNDTLAND

Chairman of the World Commission on Environment and Development

Oxford   New York

OXFORD UNIVERSITY PRESS

1990

Oxford University Press, Walton Street, Oxford OX2 6DP

Oxford New York Toronto
Delhi Bombay Calcutta Madras Karachi
Petaling Jaya Singapore Hong Kong Tokyo
Nairobi Dar es Salaam Cape Town
Melbourne Auckland
and associated companies in
Berlin Ibadan

Oxford is a trade mark of Oxford University Press

British Library Cataloguing in Publication Data
Starke, Linda
Signs of hope: working towards our common future.
1. Environment. Conservation
I. Title
333.72
ISBN 0-19-212993-7
ISBN 0-19-285225-6 (pbk)

Library of Congress Cataloging in Publication Data
Data available

Printed on recycled paper

Set by Oxford Text System
Printed in Great Britain by
Clays Ltd.
Bungay, Suffolk

Contents

List of Boxes

Preface

In August 1989 I received a telephone call from Thomas Webster of Oxford University Press. Thomas had worked very closely with the World Commission on Environment and Development in 1987 on the publication of its final report, *Our Common Future*. He noted that the report was still selling very well around the world and that from his review of the *Brundtland Bulletins* it was obvious that it was having a significant impact. Would this not, he asked, be a good time to take a look at the impact of the report, and at changes in the world regarding environment and development issues since 1987?

As one of the objectives of the Centre for Our Common Future is to track these changes, I certainly knew that a great deal had happened over the last three years, and especially in the last twelve months. Indeed, the Centre is inundated with information about global, national, and local developments, as reflected in the expansion of our quarterly *Brundtland Bulletin* to 120 pages by December 1989.

We agreed that an update on progress towards our common future could make a contribution to the unfolding international dialogue on these issues. To write the book, I turned to Linda Starke, who worked with the Brundtland Commission during the eight months preceding the report's publication. As editor of *Our Common Future* and also of all seven editions of Worldwatch Institute's annual *State of the World*, Linda knows well the issues the Commission covered and many of the key people working on them around the world. Her willingness to undertake this project at short notice and with a seemingly impossible deadline is greatly appreciated.

To reflect one of the underlying principles in *Our Common Future*, that the issues of agricultural development, poverty alleviation, energy use, population growth, and so on cannot be addressed in isolation, we decided to look at developments in the various sectors of society trying to address these problems: international organizations, governments, community development and environmental groups, industry, and the media. The organization of the book into these chapters should not be taken to mean that these groups can or should work in isolation either. In the nineties, in what is being called the crucial decade, all sectors of society must figure out how to work together on all the problems covered by the terms 'environment' and 'development', not one sector or one problem at a time.

Early in the project we decided to do our best to have *Signs of Hope* available in time for the May 1990 follow-up conference on the Brundtland report in Bergen, Norway, which is sponsored by the UN Economic Commission for Europe (ECE) and the government of Norway. This tight deadline unfortunately proved to be a constraint on our ability to include as much information on developments in the Third World as we would have liked. The inclusion of considerable information on Europe and North America is therefore more a reflection of our initial audience and the time constraints we were operating under than an indication of the balance of signs of hope worldwide over the last three years.

Indeed, *Signs of Hope* is not meant to be a comprehensive review of developments since *Our Common Future* was published, for so many changes have occurred around the world, and often with such breath-taking speed, that no single volume could capture them. Rather, this book is a snapshot of some of the progress that has been made towards our common future as the new decade opens, with an emphasis on developments in the ECE region. If funding for another follow-up report can be found, we would hope to look more closely at developments in the rest of world. In fact, it is my hope that the Centre can become a focal point for reports on progress worldwide, and thus be able to share the signs of hope with readers everywhere.

The other decision taken from the start was to accentuate the positive. *Signs of Hope* should be seen as complementing the valuable publications of the International Institute for Environment and Development, UN Development Programme, UN Environment Programme, UNICEF, World Resources Institute, Worldwatch, and others, all of which document the escalating pressure on the world's resource base and the worsening standard of living of most people in the world.

Janet Hunt in Australia, asked to comment on signs of hope, reassured us that we were on the right track: 'Good luck with your book. Something hopeful and positive is needed to keep up the feeling that people can make changes happen. But at the same time we must not gloss over the hard questions, which most Governments would prefer to forget about.'

Indeed, it must be said that when looking at progress that has been made to date, one is struck by how many signs of hope there are on the environment side of the equation and how few, if any, are to be found on the development side. A few of the hard questions about development are raised in Chapters 7 and 8—How do we finance sustainable development? What institutional changes are needed to address these cross-cutting problems? What must we do to launch the worldwide movement to alleviate poverty that is so needed if the new rush to save the earth is to accomplish its goals?

These questions, and several others, constitute what we have called the 'unfinished agenda'. The world's answers will determine our success or failure in securing our common future.

Warren H. Lindner
Executive Director
Centre for Our Common Future

Acknowledgements

Writing a book on a short deadline, especially when it is your first, requires quick answers and a great deal of understanding from family and friends. For supplying materials and answers at short notice, thanks go to Janet Edwards, Anya Halle, Ruth Jones, Danielle Moskal, Mary Paden, Kathy Rankin, and the staff at several embassies in Washington. For moral support, I am indebted to Steve Behrens, Karon Brashares, Kathleen Courrier, Kevin Finneran, Kim Griffith, Paul Mansfield, Gail Ross, Richard Smith, Allerd Stikker, and Ted Wolf.

Also of great help were my sometimes-colleagues at Worldwatch, especially Alan Durning, Christopher Flavin, Hilary French, Jodi Jabobson, Nicholas Lenssen, and Sandra Postel, some of whom I pestered with questions right up to the last minute; an award for proof-reading above and beyond the call of duty goes to Lester Brown.

Special thanks to Beatrice Olivastri for being the first person in the real world to read any chapters, and for telling me to stop worrying. And to my parents, Joe and Joanna Starke, for letting me put a book deadline ahead of my need to spend time with them.

Signs of Hope includes quotes from a number of people who responded to the Centre for Our Common Future's request for comments on signs of hope and the unfinished agenda. Putting together a picture of where we are and where we need to be would have been impossible without these letters from Michael Gucovsky, Janet Hunt, Ashok Khosla, Robert Lamb, Mochtar Lubis, Ken Piddington, Shridath Ramphal, Margaret Thatcher, Jon Tinker, and Mostafa Tolba. I am especially grateful to those who took time from busy schedules to sit down with me and talk about the book: Federal Deputy Fabio Feldman of Brazil, Beth Goodson and Borje Svensson of the IFPAAW in Geneva, Ira Kaufman of Legacy International, Diana Page of WRI, former Commissioners Bill Ruckelshaus and Maurice Strong, and Lloyd Timberlake of IIED.

The people I worked with at Oxford University Press on *Our Common Future* are still able to turn around a manuscript in record time, and to stay in a good humour while doing it. My thanks to Cyril Cox, who let me get away with *some* Americanisms; to Graham Roberts and the OTS production staff; and to Thomas Webster, without whom this book would not exist.

That it does exist, and that it could be done so quickly, is due to the Centre for Our Common Future; my appreciation for their help goes to all the staff there. I knew last year that my views on recent signs of hope and on the unfinished agenda meshed with those of the Centre. I did not know I would be able to write a book in twelve weeks on those topics. For giving me the opportunity to do so, and for never doubting I could, I am very grateful to Chip Lindner, the indefatigable Executive Director of the Centre.

When he asked me to write this book, Chip Lindner said his aim was 'not to tell people that everything is awful, but that everything is possible'. This book is written in the spirit of the possible.

Linda Starke
Washington, DC

Foreword

Three years ago, when *Our Common Future*, the report of the World Commission on Environment and Development, was first published, its positive, even optimistic tone took many people by surprise. In declining merely to add to the growing pile of documents warning of doom, my twenty-one fellow Commissioners and I chose instead to fulfil the mandate we had received from the General Assembly of the United Nations. Among other charges, we were to formulate 'innovative, concrete and realistic action proposals' with which to confront the critical issues of environment and development.

Our Common Future is a hard-won consensus of policy principles forming the basis for sound and responsible management of the Earth's resources and the common future of all its creatures. That a majority of world governments, all the major international institutions, and key non-governmental organizations have since accepted and endorsed our findings as the basis for future policy-making is a dramatic testimony to a common recognition of the new realities, and reflects a worldwide desire to find the co-operative route forward.

The report has had a profound and growing influence on the world's makers and shapers of policy and of public opinion. It has been debated and commented upon, and the issues it deals with have been taken up through the mass media, popularized and disseminated throughout the world.

Unpredicted, towards the end of the eighties, an extraordinary shift started to take place in the international scene. The Cold War, which had hung like a dark cloud over us all for more than forty years, was suddenly at an end. The superpower confrontation crumbled away. Old nations reawakened. Relationships within and among nations suddenly, almost overnight, altered fundamentally. That most of what occurred did so peacefully is all the more remarkable. By the beginning of the nineties, the northern hemisphere at least seemed set towards a radically new political destiny and an era of change and new possibilities.

In the southern world, by contrast, little seems to have changed. Here, in the developing countries where the overwhelming majority of human beings live, and where perhaps a quarter of the world's people go to bed hungry and without hope each night, the vast problems of mass poverty and its consequences seem only to worsen, barely touched by events elsewhere. The hopes of a better future that have arisen in the north have yet to touch the poorest nations.

The impoverishment inflicted on the southern world, in large part by an economic disequilibrium imposed with neither reason nor foresight by northern consumer society, must go. It is both a chief cause and a chief effect of a massive and growing environmental destruction that ultimately threatens all of us. Furthermore, it has a human cost that contains the seeds of future instability and conflict, and is morally repugnant.

Yet there are grounds for hope, not least because of a sense that is growing fast in the undercurrents of collective human awareness, as people everywhere come to grasp the scale and imminence of the real, overwhelming threats to our survival. This phenomenon is occurring among people in prosperous industrial societies as much as among the poor of the Earth. It knows no political, economic, national, or ethnic boundaries. There is an almost universal perception that a whole new complex of realities is upon us as a species, and that no stockpile of weapons can protect us from the new threats to common survival—that, in short, we need a common, shared, range of solutions.

Few thinking people today would question that a new international political and economic dispensation must now be brought into being within the next few years if we are to turn a rapidly emerging crisis aside and bring a new era of hope to realization. Few would now question that only through profound adaptation can Humanity suvive at all, or that only through the willing, self-interested participation of all nations and all people can we restore health to our world and hope to Humanity. The challenge of the nineties is to act with rapidity and boldness in bringing the desired changes into being so that, by the turn of the millennium, and no later, our common future will have been secured. Today, this is a feasible objective.

We have much in our favour. The readiness of the superpowers to enter a new era of co-operation is a major factor. The present relative degree of political stability in the world is another, although we dare never allow ourselves to abandon watchfulness for potential conflicts. Our scientific, industrial, and technological resources are vast, our institutional mechanisms are capable of being developed to manage new patterns of international co-operation and to preserve peace. Worldwide communications networks mean that no one need be unaware of changing global realities. A comparison of the arithmetic of the now anachronistic levels of arms expenditure with the sums required to eradicate world poverty has exposed the full disgrace of our failure to do so.

Downstream of the debate, the great international conferences of coming years offer a real opportunity to convert those policy principles, the 'strategic imperatives' our Commission laid down, into international treaties that will form the basis of the new international responsibility human beings increasingly will demand, and on which the world's future without question depends. Last year's upsurge of popular willpower, bringing about changes in the world unthinkable only months earlier, showed that it is possible to break with old patterns and awaken new hopes.

I welcome the publication of *Signs of Hope*, which will add force and direction, and encouragement, to this growing momentum for change.

Gro Harlem Brundtland
Oslo, March 1990

1

'Our Common Future'

Moscow, December 1986. A group of people gather around a hotel conference table to consider the nearly final draft of a report on environment and development they have worked on for three years. As discussion of the first chapter opens, someone suggests adding a question mark to the working title of 'A Threatened Future'. The debate on that idea does not last long. All assembled agree, with good cause, that our future is undeniably threatened.

It is even more threatened now. But over the last three years the world has started to pay a little attention—not enough, but a little—to some of the sources of those threats. And to take a few halting steps towards a society based on sustainable development.

*   *   *

The twenty-two people who met that cold day in Moscow constituted the World Commission on Environment and Development. The report it published five months later was *Our Common Future*, and the patterns of development it documented left no question that our future is threatened. Yet the overall message of the report is that the world can change, that 'people can build a future that is more prosperous, more just, and more secure'.

The world has indeed changed since the Commission's report was published. In some cases, *Our Common Future* has provided governments, private groups, and individuals with a powerful tool that helped them reassess where they were headed, and rethink their approach to sharing this only one earth. But the world has also changed because the problems the Commission wrote about— deforestation, poverty, climate change, species loss, the debt crisis, depletion of the ozone layer, and so on—have become more obvious, and all the more pressing.

The pace of change, the speed with which we humans are altering our physical environment, is without precedent. The signs of stress are well documented elsewhere: stress to the earth's natural support systems and to the world's political, economic, and social support systems. Fortunately, the speed with which some sectors of society are starting to react to these alterations is also notable. Thus, a report on progress towards our common future over the last three years seems in order.

In large part, the signs of hope this book documents are about the world's new-found environmentalism: Rock stars and schoolchildren sing about the tragedy of the disappearing rain forests. Governments pledge to protect the earth's ozone layer. Three out of four Americans identify themselves as environmentalists. Mayors and ministers of energy launch tree-planting campaigns in a symbolic attempt to slow global warming. Green consumers make their voices heard at the check-out counter, and manufacturers respond with 'future-friendly' products. Contrary to the predictions of sceptics who see all this as a passing fad, public and political attention to environmental issues is not going to fade. Television screens and newspaper headlines in the nineties will not let it.

Those screens and headlines have already been filled with unforgettable scenes as the decade opens: the wall dividing Berlin coming down, and the seemingly impenetrable walls of a South African prison suddenly giving way. The connection of these joyful scenes to environment and development problems may not be immediately clear, but it is none the less real. For both these historic moments signal moves towards the most basic prerequisite for the pursuit of sustainable development: a political system that secures, in the words of the Commission, effective citizen participation in decision-making. Until this is a reality everywhere, that more prosperous, more just, more secure future cannot be built.

These few hopeful signs are offset, some would say outweighed, by the lack of progress on the other focus of the Commission's work—development, and meeting the essential needs of the world's poor. More people live in absolute poverty today than ten years ago, which is deplorable but not so surprising because of the arithmetic of population growth. Moreover, a higher proportion of us now fall into this desperate category: nearly one in four people in the world.

Some of us have started to think about how we should stop mortgaging the future of our children. But we have done little to stop squandering the present of many who share the earth with us today. And few seem to understand yet the folly of trying to do one but not the other. As the Commission pointed out, 'a failure to address the interaction between resource depletion and rising poverty will accelerate global ecological deterioration'. This failure is touched on throughout the chapters that follow, especially the final one on the unfinished agenda.

*   *   *

The World Commission on Environment and Development concluded that signs of stress were accompanied by some signs of hope after nearly three years of meeting in private sessions, of discussions with scientists and with political leaders, and of public hearings attended by thousands of people (see Box 1–1 for a summary of main recommendations). To understand the foundation for their conclusions, as well as the interest the report generated and the enthusiasm with which it was greeted, it is useful to consider the genesis of the Commission and its approach to the task it was set.

In December 1983, in response to a UN General Assembly resolution, Secretary-General Perez de Cuellar appointed Gro Harlem Brundtland as Chairman of an independent World Commission on Environment and Development, and named Dr Mansour Khalid as its Vice-Chairman. These two were asked to appoint the remaining members of the Commission; all served in their individual capacities, not as representatives of any government or institution (Appendix 1 has a list of the members).

The mandate adopted by the Brundtland Commission, as it came to be known, was threefold:

- to re-examine the critical issues of environment and development and to formulate innovative, concrete, and realistic action proposals to deal with them;
- to strengthen international co-operation on environment and development and to assess and propose new forms of co-operation that can break out of existing patterns and influence policies and events in the direction of needed change; and
- to raise the level of understanding and commitment to action on the part of individuals, voluntary organizations, businesses, institutes and governments.

Box 1–1. Main Recommendations of the World Commission on Environment and Development

In February 1987 the final meeting of the Commission was held in Tokyo. The members called upon 'all the nations of the World, both jointly and individually, to integrate sustainable development into their goals and to adopt the following principles to guide their policy actions':

- **Revive Growth**: Poverty is a major source of environmental degradation. . . . Economic growth must be stimulated, particularly in developing countries, while enhancing the resource base. The industrialized countries can and must contribute to reviving world economic growth.

- **Change the Quality of Growth**: Revived growth must be of a new kind in which sustainability, equity, social justice, and security are firmly embedded as major social goals. A safe, environmentally sound energy pathway is an indispensable component of this. . . . Better income distribution, reduced vulnerability to natural disasters and technological risks, improved health, and preservation of cultural heritage—all contribute to raising the quality of that growth.

- **Conserve and Enhance the Resource Base**: Sustainability requires the conservation of environmental resources such as clean air, water, forests, and soils; maintaining genetic diversity; and using energy, water, and raw materials efficiently. Improvements in the efficiency of production must be accelerated to reduce per capita consumption of natural resources and encourage a shift to non-polluting products and technologies.

- **Ensure a Sustainable Level of Population**: Population policies should be formulated and integrated with other economic and social development programmes—education, health care, and the expansion of the livelihood base of the poor.

Extending the work of commissions that produced reports in the early eighties on development and disarmament, chaired respectively by Willy Brandt and Olof Palme, the Brundtland Commission decided to both study the issues and listen to people who are experts by virtue of their daily struggles. Public hearings were held on all five continents. The Commissioners wanted to hear firsthand about the pressures people face as they try to meet food, housing, clothing, health care, and education needs—and of the pressures they in turn place on natural resources they know full well are all they have to rely on.

- **Reorient Technology and Manage Risks**: The capacity for technological innovation needs to be greatly enhanced in developing countries. The orientation of technology development in all countries must also be changed to pay greater regard to environmental factors. . . . Greater public participation and free access to relevant information should be promoted in decision-making processes touching on environment and development issues.

- **Integrate Environment and Economics in Decision-making**: Sustainability requires the enforcement of wider responsibilities for the impacts of policy decisions. Those making such policy decisions must be responsible for the impact of those decisions upon the environmental resource capital of their nations. They must focus on the sources of environmental damage rather than the symptoms.

- **Reform International Economic Relations**: Fundamental improvements in market access, technology transfer, and international finance are necessary to help developing countries widen their opportunities by diversifying their economic and trade bases and building their self-reliance.

- **Strengthen International Co-operation**: Higher priorities must be assigned to environmental monitoring, assessment, research and development, and resource management in all fields of international development. This requires a high level of commitment by all countries to the satisfactory working of multilateral institutions; to the making and observance of international rules in fields such as trade and investment; and to constructive dialogue on the many issues where national interests do not immediately coincide. . . . New dimensions of multilateralism are essential to sustainable human progress.

Source: Excerpted from the Tokyo Declaration, World Commission on Environment and Development, 27 February 1987.

These meetings with the public—people who were always interested in the Commission's work, and sometimes critical of it—without a doubt helped the Commissioners to develop their thinking on these complex issues. No one in the normal course of a lifetime is exposed to the concerns of indigenous people in Brazil and Canada, the tales of farmers in Indonesia and the Soviet Union, the opinions of environmentalists in Oslo and Harare, and the views of Presidents and Prime Ministers around the world. So each Commissioner had something to learn. In the closed-door sessions where the substance of the final report was discussed, Commissioners often recalled the testimony of the everyday experts, the people who spoke from the heart about their struggles.

The issues addressed during the meetings were of great interest wherever the Commission went, but the stakes were raised even more in early 1986, when the Chairman, Mrs Brundtland, became Prime Minister of Norway. This guaranteed much greater media and public awareness of the Commission and its work, and provided a credibility to subsequent public hearings that would not otherwise have been possible.

The net result of the Commission's open approach was more than 10,000 pages of transcripts and written submissions from hundreds of organizations and individuals. It also gave a real-world feeling to the final report, reflected in excerpts from the hearings that appear throughout *Our Common Future*. And it meant that the report was eagerly awaited by the many people who had testified at or attended hearings in Jakarta, Oslo, São Paulo, Brasilia, Harare, Nairobi, Moscow, Tokyo, and six Canadian cities.

Their wait ended on 27 April 1987. To a capacity crowd in the Queen Elizabeth Hall in London, Prime Minister Brundtland released the report, 'to make the world aware that humanity has come to a crossroad'. In doing so, she delivered it to twelve young people from around the world who had testified before the Commission or helped it during meetings in their countries. This involvement of young people in the launch of *Our Common Future* followed through on part of the Commission's original mandate: to seek especially the views of youth. Before handing each of them a copy of the book, the Chairman said:

> Securing our common future will require new energy and openness, fresh insights, and an ability to look beyond the narrow bounds of national frontiers and separate scientific disciplines. The young are better at such vision than we, who are too often constrained by the traditions of a former, more fragmented world. We must tap their energy, their openness, their ability to see the interdependence of issues. . . .

> Our generation has too often been willing to use the resources of the future to meet our own short-term goals. It is a debt we can never repay. If we fail to change our ways, these young men and women will suffer more than we, and they and their children will be denied their fundamental right to a healthy, productive, life-enhancing environment.

Breaking with the traditional role of bodies such as these, the Commission was determined not to have this be the end of their involvement, to let the report gather dust on policy-makers' shelves.

In the few months left between the launch in April and the report's scheduled presentation to the UN General Assembly that October, *Our Common Future* was on the agenda at numerous governmental and UN meetings. Before it even reached the floor of the United Nations, a National Task Force in Canada brought out a response to the report, and a reader's guide was published by the International Institute for Environment and Development in London.

In December 1987 a resolution by the General Assembly called on all governments and governing bodies of organizations, bodies, and programmes within the UN family to report in September 1989 on progress made towards achieving environmentally sound and sustainable development. Thus the wheels were set in motion to use the Commission's report as the basis of reviewing existing programmes as well as those on the drawing-boards.

By mid-March 1989, twenty-two governments and all UN organizations had submitted reports to Secretary-General de Cuellar that illustrated a broad acceptance of the principles in *Our Common Future* and a wide range of efforts to incorporate those principles in programmes around the world. Outside the UN system, the report also became a powerful political tool for local and national environment and development groups that joined together and formed their own responses. Numerous groups have even sprung up around the world with 'our common future' in their titles, and the report as their rallying point.

Two years after its presentation to the United Nations that October, an estimated half-million copies of the report were in print in Arabic, Bulgarian, Chinese, Danish, English, Finnish, French, German, Hindi, Hungarian, Indonesian, Italian, Japanese, Norwegian, Portuguese, Russian, Spanish, Swedish, and Turkish. Versions in Czech, Dutch, Hebrew, Polish, Sinhala, and Urdu are expected soon. Whether as *Notre Avenir a Tous*, *Hari Depan Kita Bertsama*, or *Vores Faelles Fremtid*, the Commission's report has had an impact far greater than anyone at the launch could have imagined.

Our Common Future struck a chord of public concern when it was published, and people's willingness to consider the conclusions of the Commission reflect that concern. Some of the report's impact also stems from the existence of the Centre for Our Common Future, which opened its doors one year after the launch of the report. Established to maintain the momentum created by the publication of *Our Common Future*, the Centre has about 150 Working Partners around the world who join it in trying to build a broad base of support for sustainable development (see Appendix 1 for a fuller description of the Centre's work.)

As the focal point for anyone wishing to follow up on the work of the Brundtland Commission, the Centre receives daily, indeed is swamped with, requests for information as well as reports of related activities in country after country. It was the impressive amount of these activities that led the Centre to undertake this book, to review progress in the many areas the Commission discussed and to point out where progress is still lacking.

Reviewing the files at the Centre for a few weeks gives the impression that everyone in the world is talking about *Our Common Future*, and that half of them are about to catch a plane to attend yet another vital meeting on the topic. This really is not the case. But many dedicated people around the world are indeed working on their own visions of a sustainable future. Out of this chaos, our common future will be forged.

*   *   *

At the core of *Our Common Future* lies the principle of sustainable development, a key concept for evaluating the signs of hope discussed in this book. The Brundtland Commission defined sustainable development as meeting 'the needs of the present without compromising the ability of future generations to meet their own needs'. It is a very simple definition, and its vagueness has been hailed by some as its strength. Indeed, the Commission went on to point out that 'no single blueprint of sustainability will be found, as economic and social systems and ecological conditions differ widely among countries. Each nation will have to work out its own concrete policy implications.'

Since 1987, a handful of nations have undertaken that task. And the few words defining sustainable development have become the starting-point for countless debates about its implications for various sectors of society. Yet all can understand one basic point: over-use of the resources they depend on now will leave less for their children and grandchildren to draw on.

The Commission's embrace of sustainable development as an underlying principle gave political credibility to a concept many others had worked on over the previous decade. It is unclear who coined the term, but by 1980 it was enshrined in the title of a key document for the eighties—the *World Conservation Strategy: Living Resource Conservation for Sustainable Development*, published by the International Union for Conservation of Nature and Natural Resources, the World Wildlife Fund, and the UN Environment Programme. That *Strategy*'s definition has stood the test of time well: 'For development to be sustainable it must take account of social and ecological factors, as well as economic ones; of the living and non-living resource base; and of the long term as well as the short term advantages and disadvantages of alternative actions.'

Some people claim that sustainable development is just the newest weapon in a campaign by western industrial nations to dictate the economic policies the Third World must follow. They are concerned that lending from multilateral development banks will be restricted to projects judged 'sustainable', and will be taken away from projects the countries themselves believe need support. But the Commission strongly rejected this approach: 'It is vitally important for development that there should be a substantial increase in resources available to the World Bank and IDA [the International Development Association]. Increased commercial bank lending is also necessary for major debtors.' At the same time, the report noted that 'more aid and other forms of finance, while necessary, are not sufficient. Projects and programmes must be designed for sustainable development.'

A second key principle in *Our Common Future* is the inter-disciplinary nature of the world's environment and development problems as well as their solutions. 'Sectoral organizations', the Commission pointed out, 'tend to pursue sectoral objectives and to treat their impacts on other sectors as side effects, taken into account only if compelled to do so. . . . Many of the environment and development problems that confront us have their roots in this sectoral fragmentation of responsibility.'

The organization of this book reflects that principle. Rather than looking at specific issues addressed by the Commission—food supplies or patterns of energy use, for example—*Signs of Hope* looks at recent efforts by international bodies, governments, citizens' groups, industry, and the media to consider all the various issues. Many of the hopeful signs stem from new ways of thinking about our common future, not from concrete policy changes to date. So far, these have been too few.

Even this division into chapters on the various agents of change for our common future is of course artificial. In Canada, for instance, a National Round Table on the Environment and the Economy and various provincial Round Tables bring many of these people together, in a direct response to the Commission's call to action. Indeed, many things are happening in Canada that cut across traditional sectors, so efforts in that nation will come up time and again. They will be put into one chapter or another, as seems appropriate, but bear in mind that in an ideal world all the players would be at the table, working together towards the goal of development that can endure.

A handful of the many people working for change around the world, including former Commissioners and others associated with its work, were invited to reflect on how their views of the world's prospects have altered in the last few years. They were also asked to name the most important items on the global agenda for the next three to five years. Their responses are incorporated throughout this book.

*   *   *

The opening pages of *Our Common Future* included a list of events during the 900 days between the Commission's first meeting and the publication of the report: the Bhopal disaster . . . Chernobyl . . . the drought in Africa that put thirty-five million lives at risk . . . the Mexico City gas-tank explosion . . . the accidental release of chemicals into the Rhine during a fire. This recitation has been reprinted many times and in many places when the work of the Commission is discussed. It must be a basic human instinct to want to count things. Perhaps it gives us the illusion that we are more in control of events.

The list that can be drawn up today is as daunting. In the 900 days since the report's publication, twenty-five million people were left homeless in Bangladesh by flooding that was aided and abetted by deforestation in surrounding hillsides . . . oil spills in Alaska and California changed vulnerable and unique ecosystems, perhaps forever . . . on a sweltering June day, a leading government scientist told the US Senate that global warming had begun . . . other scientists announced that two things seemed much worse than models had predicted: the hole in the ozone layer and the rate of

loss of tropical rain forests . . . twenty-two million children under
the age of 5 died from treatable, preventable diseases.

At the same time, however, the following news headlines from
the last three years would have been unthinkable in 1987: an
economic summit of the seven leading industrial countries that
devoted one third of its communiqué to concerns about the
environment . . . the suspension of commercial logging permits by
the Brazilian and Thai governments . . . a reversal of the UK
government's commitment to subject nuclear power to the rigour
of the private market . . . the free distribution in Africa of a drug
that might eliminate river blindness . . . a comprehensive proposal
in southern California to clean up the air . . . an accord signed in
Montreal to halve the production of chemicals depleting the earth's
protective ozone layer. And so on and so on.

This book is about that list. Is enough being done? No. Is more
being done than just three years ago? Unequivocally yes. We are
living in a time of enormous opportunity—an opportunity matched
in scale only by that of the challenge we face. The prospects for
seizing that opportunity have brightened a little since 1987, when
the Commission issued a call for action.

People also love to make lists of 'how to save the planet'. The
last few years have seen a resurgence of these in response to
consumer interest, ranging from the US Greens' '101 Things You
Can Do To Promote Green Values' to a book called *Save Our Planet:
750 Everyday Ways You Can Help Clean Up the Earth*. These lists
are important for helping people see that they can have an impact,
that their day-to-day decisions affect the prospects for sustainable
development.

Signs of Hope provides no such list, although the final chapter
does focus on items on the unfinished agenda for sustainable
development world-wide. But one list applicable to all who read
these words is certainly not possible. There are many paths to
sustainable development. Reading about some of the ones that
people have started down may at least provide a feeling that the
foundation for the future is being put in place.

Our common future is being shaped in literally countless ways
by people all over the world. Some of them make headlines. A very
few of them are heard in the chapters that follow. But the
overwhelming majority are people who do not make the news.
They make the difference.

2

Global Responses to Global Problems

In December 1989 the front pages of the world's newspapers and the evening television newscasts captured another of the historic scenes of the last twelve months: the Presidents of the Soviet Union and the United States giving a joint press conference to describe the success of their meetings in Malta. With pledges to reduce conventional armed forces in Europe, to halve US and Soviet strategic nuclear arsenals, and to cut chemical weapon stockpiles drastically, the only rough waters seemed to be those of the Mediterranean. And the only ill feelings between the two major powers seemed to be over a cancelled dinner party on the *Maxim Gorkii*.

As the Brundtland Commission pointed out, 'among the dangers facing the environment, the possibility of nuclear war, or military conflict of a lesser scale involving weapons of mass destruction, is undoubtedly the gravest'. Recent progress towards disarmament pulls the world a few steps further back from the brink.

It also helps turn attention to the next point made by the Commission:

> A comprehensive approach to international and national security must transcend the traditional emphasis on military power and armed competition. The real sources of insecurity also encompass unsustainable development. . . . There are, of course, no military solutions to 'environmental insecurity.' And modern warfare can itself create major internationally shared environmental hazards.

Furthermore, the idea of national sovereignty has been fundamentally modified by the fact of interdependence in the realm of economics, environment, and security. The global commons cannot be managed from any national centre: The nation state is insufficient to deal with threats to shared ecosystems. Threats to environmental security can only be dealt with by joint management and multilateral procedures and mechanisms.

Presidents Bush and Gorbachev talked about more than just disarmament in Malta: Mr Bush offered to host an international conference in Washington in late 1990 to begin negotiations on a treaty about global warming. That such an issue would be brought to the table at a summit—or even a mini-summit, as it was labelled—is a striking indication of the new era of ecological diplomacy the world has entered.

A few months before this historic meeting, George Bush joined the leaders of the other six large industrial countries in Paris for their annual economic summit. In acknowledging that development is 'a shared global challenge', the Group of Seven pledged to help developing countries by opening the world trading system, welcomed the decision of some governments to 'write off' outstanding loans to thirteen of the poorest countries, and urged each other to meet the targets set for official development assistance. All this was nothing new, however. Such pledges have been heard before; subsequent actions too seldom match the words.

What was new this time was the attention given specifically to the environment. At the 1988 summit, in Toronto, the final statement contained three paragraphs on the environment, and noted that 'we endorse the concept of sustainable development'. A year later, one third of the fifty-six paragraphs in the final communiqué had this heading (see Box 2–1).

The Group of Seven leaders' prescription for sustainable development made passing reference to the connection of environmental protection to the other issues they addressed. But the separate section on 'environment' and the media's focus on the 'green' summit indicate that the link between economy and ecology, between pervasive poverty and environmental degradation, is not yet clear. As the Brundtland Commission noted, 'failure to address the interaction between resource depletion and rising poverty will accelerate global ecological deterioration'.

The symptoms of global environmental stress that spurred the Group of Seven to devote so much time and space to the earth are by now depressingly familiar: the depletion of the protective ozone layer, brought to the world's attention by the 'hole' over

Antarctica . . . global warming, also called climate change or the 'greenhouse effect' . . . the involuntary pollution of the air, water, and soils of neighbouring countries, and the intentional dumping of hazardous wastes in some cases . . . the loss of tropical forests, and with them countless species of plants and animals.

Recent moves towards disarmament could mean the military needs a smaller portion of national budgets, and perhaps these pressing environmental problems can receive a larger share. Newspapers are full of talk of such a 'peace dividend', but the line of government agencies and private causes wanting a share of it is long.

Soviet Foreign Minister Eduard Shevardnadze, writing in *Literaturnaya Gazeta*, believes 'an international ecological fund, formed by deductions from resources released by cuts in military spending in the course of disarmament, is a practical and feasible possibility'. A few months later, speaking in Moscow in January 1990, Gro Harlem Brundtland noted that 'as East and West move from confrontation through dialogue to cooperation, enormous human and financial resources will be unleashed. As the physical barriers between us come down, we must build together a new coalition of reason, a coalition for our common security and our common survival.'

In the meantime, while petitioners fight over the spoils of peace, in response to these global problems international diplomats have been doing the detailed and difficult work of writing new treaties and redrafting old ones. This chapter discusses recent significant initiatives on chlorofluorocarbons (CFCs), global warming, forestry, species loss, biological diversity, and the way development projects are designed. But it is worth noting that the chapter is entitled 'global responses' because, sadly, too few of the initiatives could be called 'solutions'.

In addition, the small signs of hope documented here deal principally with the physical resources of the earth. Little good news can be reported about even a global response to the need to eliminate poverty, let alone a solution.

*　　*　　*

One issue is universally cited as a stunning example of international co-operation to address an environmental problem: the depletion of the ozone layer, caused predominantly by a class of chemicals

Box 2–1.   The 'Green' Summit of Industrial Nations

In July 1989 the leaders of Canada, France, Italy, Japan, the United Kingdom, the United States, and West Germany met for their fifteenth annual economic summit. Environmental threats, they acknowledged, deserve as much attention as economic ones:

■ There is growing awareness throughout the world of the necessity to preserve better the global ecological balance. This includes serious threats to the atmosphere, which could lead to future climate changes. We note with great concern the growing pollution of air, lakes, rivers, oceans and seas; acid rain, dangerous substances; and the rapid desertification and deforestation. Such environmental degradation endangers species and undermines the well-being of individuals and societies.

■ Decisive action is urgently needed to understand and protect the earth's ecological balance. We will work together to achieve the common goals of preserving a healthy and balanced global environment in order to meet shared economic and social objectives and to carry out obligations to future generations. . . .

■ The persisting uncertainty on some of these issues should not unduly delay our action. In this connection, we ask all countries to combine their efforts in order to improve observation and monitoring on a global scale. . . .

■ Environmental protection is integral to issues such as trade, development, energy, transport, agriculture and economic planning. Therefore, environmental considerations must be taken into account in economic decision-making. In fact good economic policies and good environmental policies are mutually reinforcing. In order to achieve sustainable development, we shall ensure the compatibility of economic growth and development with the protection of the environment. Environmental protection and related investment should contribute to economic growth. . . .

called chlorofluorocarbons. Ozone in the upper atmosphere stops harmful ultraviolet radiation from reaching the earth; scientists pointed out that the thinning of this protective shield could raise skin-cancer and cataract rates, reduce crop yields, and harm marine organisms close to the ocean's surface. In the mid-eighties, they documented that chlorine atoms set loose from CFCs in the stratosphere in a complex chemical chain reaction were eating a hole in the ozone layer over Antarctica.

US Interior Secretary Don Hodel's response to this rapidly unfolding problem was to recommend hats and sun-glasses for

- To help developing countries deal with past damage and to encourage them to take environmentally desirable action, economic incentives may include the use of aid mechanisms and specific transfer of technology. In special cases, ODA [official development assistance] debt forgiveness and debt for nature swaps can play a useful role in environmental protection. We also emphasize the necessity to take into account the interests and needs of developing countries in sustaining the growth of their economies and the financial and technological requirements to meet environmental challenges. . . .

- We strongly advocate common efforts to limit emissions of carbon dioxide and other greenhouse gases, which threaten to induce climate change, endangering the environment and ultimately the economy. . . .

- The increasing complexity of the issues related to the protection of the atmosphere calls for innovative solutions. New instruments may be contemplated. We believe that the conclusion of a framework or umbrella convention on climate change to set out general principles or guidelines is urgently required to mobilize and rationalize the efforts made by the international community. . . .

- We advocate that existing environment institutions be strengthened with the United Nations system. In particular, the United Nations Environment Program urgently requires strengthening and increased financial support. Some of us have agreed that the establishment within the United Nations of a new institution may also be worth considering.

Source: Excerpted from 'Economic Declaration', Summit of the Arch, Paris, 16 July 1989.

everyone. Fortunately, cooler heads prevailed. In 1985 the Vienna Convention for the Protection of the Ozone Layer was adopted as an initial attempt to address this rapidly unfolding problem. Richard Benedick, then chief US negotiator on the issue, points out that the Vienna Convention 'might serve as a model for those negotiating similar issues in future. By allowing nations to agree that a problem existed without immediately deciding how to deal with it, the convention paved the way for further negotiations.'

This ground-breaking step was orchestrated by the UN Environment Programme (UNEP), which by September 1987 brought government representatives together in Montreal to consider a Protocol on Substances that Deplete the Ozone Layer. Twenty-four governments and the European Community Commission indicated a willingness to freeze the production of CFCs in their countries at its 1986 level by 1989, and to halve production by 1998. The production of halons, a chemical used in fire-extinguishers and also implicated in ozone depletion, was set to halt at 1986 levels by 1992. Acknowledging the differing responsibilities for this global problem, developing countries with low per capita use of CFCs were given a ten-year grace period before the provisions of the treaty would apply.

The Montreal Protocol entered into effect in January 1989 after governments of countries representing two thirds of global use had ratified it. It is significant for two steps governments agreed to take: protect the environment before all the evidence was in on how bad damage could or would be, and make industry change its ways before alternative processes were known.

Scientists soon found evidence of escalating damage, and the resulting pressure to revise the treaty before the ink was barely dry illustrates the new partnerships that must be forged in the nineties. As Mrs Brundtland put it in a May 1989 speech:

> The scientist's chair is now firmly drawn up to the negotiating table, right next to that of the politician, the corporate manager, the lawyer, the economist and the civic leader. Indeed, moving beyond compartmentalization and outmoded patterns to draw upon the very best of our intellectual and moral resources from every field of endeavor lies at the very heart of the concept of sustainable development.

Current efforts to renegotiate the Montreal Protocol are widely expected to lead to a tighter deadline for the complete ban on CFC production. In the meantime, numerous governments have introduced their own accelerated phase-out targets (see Chapter 3). And the British government joined UNEP in hosting a March 1989 'Saving the Ozone Conference' in London, attended by 123 nations. Although Prime Minister Thatcher's goal was to get more governments to sign the Montreal Accord, Environment Minister Nicholas Ridley reported that 'the provisions of the Montreal Protocol were criticised by many delegations as not going far enough. There was general acceptance that the ultimate objective has to be the total elimination of production and consumption of CFCs and halons.'

The London meeting also provided an occasion for developing countries to raise the difficult issue of how dependent they already are on CFCs, these 'chemical wonders' that industrial nations had been promoting for so long. Turning away from the industrial technologies that use CFCs, they stressed, would require financial and technical assistance. As the Conference Chairman, Minister Ridley noted that 'ways of helping developing countries should be a major feature of the review of the Protocol, and should be urgently examined in all appropriate international contexts'.

The issue was taken up weeks later at the first meeting of Montreal Protocol signatories, in Helsinki, where a non-binding declaration called for an end to CFC production by the year 2000 at the latest. Industrial countries indicated some willingness to help the Third World find and fund alternative technologies and chemical substitutes for CFCs. But the details of where the money would actually come from remain up in the air. (See Chapter 7 for further discussion of financing proposals.)

To avoid wasting the progress to date on this global issue, these commitments to help the Third World must be met. Commonwealth Secretary-General Shridath Ramphal, a former member of the Brundtland Commission, spoke about the dangers of not doing so at a Tokyo conference on sustainable development in late 1989:

> There remain, however, lingering doubts about whether the commitments made in London and Helsinki on financial and technical assistance to developing countries—to meet the costs of adjustment from CFCs—will be adequate. The view which most developing countries have taken, and which is surely right, is that while the agreements may not be ideal, their ratification is necessary to establish a framework of multilateral control; and they now look to developed countries to respond to this act of good faith with appropriate assistance on an imaginative scale. If that good faith is shattered, it will not come easily again.

*   *   *

The good faith of the Third World—indeed, of all governments—will be needed to address the other major atmospheric issue, global warming. Although a few scientists voice scepticism about its imminent arrival or the extent of climate change to be expected, the overwhelming majority believe the build-up of 'greenhouse gases' in the atmosphere demands immediate attention and action. In fact, the World Energy Conference—a meeting once every three years of the 'energy establishment' from nearly 100 nations, generally a conservative group—began its 1987 programme by noting, 'the existence of a greenhouse effect is not to be debated'.

The gases of concern are carbon dioxide (CO_2), nitrous oxide, methane, and CFCs; as they have built up in the atmosphere during industrialization, they have continued to allow sunlight in but made it harder for heat to escape. Computer models now predict an increase in the global average temperature in the next century of 2.5 to 5.5 degrees Celsius (4.5 to 9.9 degrees Fahrenheit). Indeed, the earth is expected to be warmer by 2030—only forty years from now—than it has been in 120,000 years.

The result would be disaster. Prime agricultural areas could suddenly find that crops will no longer thrive there. After decades of investment, irrigation systems could be useless, or in the wrong place. Tropical storms are expected to increase in number and severity. Low-lying delta regions and islands are particularly threatened by the expected global sea-level rise. In fact, the President of the Maldives, a nation of islands off the south-west coast of India, recently invited the Commonwealth Heads of Government to hold their meeting in the Maldives in 2030, but warned that it might have to be under water.

It was a 1987 speech by this President, Maumoon Abdul Gayoom, and severe flooding in Bangladesh, that brought the issue of global warming sharply to the attention of the Commonwealth. Many members of this group, which has predominantly developing countries in it, already face enormous difficulties meeting basic needs in their nations. They questioned how they could afford to adjust to climate change. Indeed, why should they be expected to foot outrageous bills for a global problem they had done little to cause?

The Commonwealth Secretary-General established an Expert Group to come up with effective, practical, and feasible protective measures that could be taken; its September 1989 report, *Climate Change: Meeting the Challenge*, submitted a plan of action that called for industrial-country assistance to the Third World on such issues as flood and cyclone warning and the collection of data on greenhouse gas emissions so that realistic targets for reductions could be set.

At the international diplomatic level, the response to global warming is being funnelled through the Intergovernmental Panel on Climate Change (IPCC). This group was set up by UNEP and the World Meteorological Organisation in 1987 to consider the scientific information rapidly becoming available, to evaluate the environmental and socioeconomic impacts of climate change, and to come up with realistic response strategies.

Based on the careful negotiations that led to the Montreal Protocol, UNEP Executive Director Mostafa Tolba expects to have a draft action plan ready in the summer of 1990. The draft will be presented at the Second World Climate Conference, scheduled for November in Geneva. After that, negotiations will probably commence on the details of a global treaty—a Law of the Atmosphere, some are calling it—that could be signed at the 1992 Conference on Environment and Development. (See Chapter 8 for discussion of the 1992 meeting.)

As with the ozone issue, the participation of the Third World in any effort to slow global warming is essential. It does no good if the United States and the Soviet Union take strong steps to lower greenhouse-gas emissions—not that they have done so yet—if China and India, for instance, are rushing to burn more coal, establish oil-burning factories, and have CFC-cooled and -insulated refrigerators in every home.

To work on this issue, the IPCC has set up a special committee 'on matters related to developing countries', with representatives of Algeria, Brazil, India, Indonesia, and Kenya joining those from France, Japan, Norway, the Soviet Union, and the United States. In June 1989 IPCC participants responded to the same concerns about finance and technology transfer that had been raised regarding CFCs. Government representatives from industrial countries pledged close to $500,000 towards developing-country involvement in the IPCC process.

The move towards a global convention received a boost in November 1989 when the government of the Netherlands hosted a two-day Ministerial Conference on Atmospheric Pollution and Climatic Change, attended by representatives of sixty-eight countries. This high-level meeting, which received a great deal of press coverage, indicates a certain frustration with the pace of international treaty negotiations. So much information is becoming available so quickly, and politicians are realizing that environmental issues are of such concern to the voters, that ministers could not wait for the next step in the IPCC process. Indeed, by early 1990 several West European governments had indicated they might go ahead and try to forge separate agreements to reduce carbon emissions in their region.

The call to action issued by the meeting in the Netherlands, known as the Noordwijk Declaration, unequivocally stated that 'stabilizing the atmospheric concentrations of greenhouse gases is an imperative goal'. The conference said a world net forest growth of 12 million hectares a year within a decade should be a provisional aim, given the role that tree growth plays in balancing atmospheric CO_2 levels. It also noted that some currently available estimates indicate the stabilization goal might entail cutting emissions from human activities by more than 50 per cent. Industrial countries were urged to consider 'the feasibility' of reducing CO_2 emissions by 20 per cent by the year 2005, a step none of them has yet dared to suggest at home (see Chapter 3).

But that day may be coming closer. A week before the meeting in Noordwijk, the US State Department co-ordinator for global warming issues, William Nitze, told the *New York Times*, 'the pressure for some sort of stabilization effort is going to be very strong and very hard to resist'. Still, rhetoric and reality have yet to mesh. At the conference itself, the United States joined Japan and the Soviet Union in making sure no specific targets were set for reducing CO_2 output. Which is why the final statement calls merely for thinking about whether such a step is feasible. Those meeting in Noordwijk said all the right things about assisting developing countries financially and technically, but as Mr Ramphal pointed out regarding CFCs, it remains to be seen if agreement can be reached on a mechanism to provide such aid.

The effort to have a climate treaty as early as possible also received a push in the right direction at a March 1989 meeting in The Hague called by the Prime Ministers of France, the Netherlands, and Norway. Seventeen heads of government attended, along with representatives of seven other nations, from industrial as well as developing countries. They issued a Declaration of The Hague, subsequently endorsed by forty governments, to rally political support around the issues of ozone depletion and global warming.

In addition to commenting on the urgent need for global initiatives, The Hague meeting was significant for calling for 'a new approach, through the development of new principles of international law including new and more effective decision-making and enforcement mechanisms'. (This aspect of The Hague Declaration is discussed in Chapter 7.) These key issues—who is going to decide what needs to be done, and how can we make nations take

steps that might be in the world's interest but not in their own?—
are at the heart of global responses to global problems.

*   *   *

People were worrying about environmental problems that cross
borders long before the ozone layer and climate change became
news, of course. For more than a decade governments have
negotiated detailed treaties on reducing emissions of the pollutants
that acidify lakes, trees, and soils. Within Europe, a Convention on
Long-Range Transboundary Air Pollution signed in 1979 signalled
governments' good intentions. Subsequent protocols signed in
Helsinki (in 1985) and Sofia (in 1988) bind participants to bring
emissions of sulphur dioxide at least 30 per cent below 1980 levels
by 1993 and to stabilize those of nitrogen oxides at 1987 levels by
1994.

As individual governments introduce even more stringent national
standards, and as people as well as pollutants continue to pour
over borders in Europe, efforts to clear the air over the continent
are likely to intensify. A similar exchange of pollutants has been a
difficult issue in the normally warm relations between Canada and
the United States. Passage of a strengthened US Clean Air Act
would certainly help establish more of a regional solution to these
pollution problems.

A more newsworthy development since the Brundtland Commission considered these issues is the nasty habit industrial countries
have of dumping their wastes in the Third World—and the global
response to this practice. Several highly publicized discoveries in
West Africa of containers leaking dangerous discarded pesticides
and other wastes from industrial nations led UNEP in February
1988 to call for an international agreement to control the transport
of such materials.

The treaty that was negotiated in fairly short order was signed
in Basle, Switzerland, in March 1989—an example of how quickly
international support can be mustered for global environmental
regulations. It gives all governments the right to prohibit import
of hazardous wastes and to control their export. Environmental
authorities in an exporting nation must have proof that wastes will
be disposed of in an environmentally sound manner before export
permission can be given. Although it has not yet entered into force,

observers are sure it will be ratified by the needed twenty governments before the end of 1990.

Commenting on the significance of the Basle Treaty, and on the
practice it controls, Dr Tolba of UNEP noted that 'our agreement
has not halted the commerce in poison, but it has signalled the
international resolve to eliminate the menace that hazardous wastes
pose to the welfare of our shared environment and to the health
of the world's peoples'. Some environmental groups consider this
treaty irrelevant, it should be noted, because of the very point Dr
Tolba makes: it does not halt the commerce in hazardous wastes;
it merely seeks to license it.

Several other environmental issues have received a great deal of
international attention recently. The problems of deforestation,
species loss, and management of Antarctica stem from the pressures
of the sum total of human activities, not from what one neighbour
does to another. They cannot be traced to the practices of any one
nation, nor can they be solved by a resolution passed by one
government.

The loss of forests, particularly in tropical regions, has been the
focus of considerable concern in recent years. Although accurate
estimates of the extent of the problem are simply not available, at
a minimum eleven million hectares of tropical forest are being
cleared annually—by large- and small-scale farmers, by cattle-
ranchers, by fuel-wood-gatherers, by road-builders, by people
looking for minerals or anything else they can wrest from the land.
Over the last thirty years, 40 per cent of these rain forests have
disappeared; worse, some experts estimate that 80 per cent will be
gone thirty years from now.

This devastation not only eliminates the habitat of countless
plant and animal species and a source of livelihood for rubber
tappers and other forest dwellers, it contributes to global warming.
Trees and soils store about three times as much carbon as the
atmosphere holds. Cutting down or burning forests releases that
carbon, and adds it to the CO_2 in the atmosphere built up through
the burning of fossil fuels in industrial countries. Worldwatch
Institute notes that deforestation probably released one fifth to half
as much carbon in 1988 as fossil-fuel burning did.

Two recent international initiatives address the problem of
deforestation. The Tropical Forestry Action Plan (TFAP) was

launched in 1985 by the UN Development Programme, the UN Food and Agriculture Organization, the World Bank, and the World Resources Institute (WRI). Through a co-ordinating office in Rome, TFAP works with national forestry departments to assess the extent of the problem and suggest target areas for remedial efforts.

The programme got off to a rocky start, by all accounts. Too little effort was made to include in the design-process all those affected by forestry issues—representatives, for example, of indigenous peoples in the Amazon. Assessments conducted in some fifty countries have generally been confined to the government-managed forestry sector, notes Tom Fox of WRI, although they have helped in revisions of development-assistance programmes in Ghana, Sudan, and Kenya. By late 1989 the TFAP Advisors Group included non-governmental organizations in its meetings. It was also considering how to affect policies outside traditional forestry departments and how to include more of those who have a stake in forests. Without this, the programme cannot succeed: effective citizen participation, the Brundtland Commission pointed out, is a prerequisite for sustainable development.

Just as the Commission was finishing its work, the International Tropical Timber Organization (ITTO) was opening its offices in Japan. This group implements a 1985 agreement on the production and use of industrial wood from tropical nations, an agreement that took nine years to negotiate. ITTO provides a sign of hope in that producing and consuming countries are working together towards the sustainable use of tropical forests; its recent initiatives to listen to the views of environmental-group representatives should increase its effectiveness. ITTO's forty-five member-countries account for 80 per cent of these forests and 95 per cent of tropical timber exports. One aim is to help producers develop better techniques for reforestation and forest management. Plans are also being made to modify logging systems in order to minimize their environmental impacts.

Efforts to halt deforestation have another important objective: slowing the loss of species and the destruction of unique ecosystems. Estimates of the size of this problem are depressing, and disconcertingly wide-ranging. According to the World Resources Institute, 5 to 15 per cent of the world's species could disappear in the next thirty years, primarily in the tropics. If there are roughly ten million species on the planet—and scientists cannot even confirm that number; some think it may be as high as thirty million—that means a loss of 50 to 150 species a day.

A long-standing international body whose sole focus is species loss is CITES, the Convention on International Trade in Endangered Species of Wild Flora and Fauna. Since 1973 this body has monitored the status of plants and animals—at least the ones scientists have identified thus far. CITES prohibits or controls trade in both live specimens and wildlife products, using a listing system that designates species as 'imminently endangered' or as 'not yet in jeopardy of extinction but being monitored'.

One of CITES's more controversial decisions was announced in October 1989, when the sale of ivory was banned and the African elephant was moved on to the endangered list. According to Greenpeace, illegal poaching over the last decade has cut the population of elephants from 1.5 million to less than 600,000. The CITES action had been preceded by unilateral bans in several nations, including the United States and Japan, on the import of ivory; the result had been a precipitous drop in ivory prices. As the effect of the ban works its way back along the chain of supply, so the theory goes, poachers will no longer be able to get $100 per kilogram of ivory.

Several African nations objected to the CITES reclassification, claiming that they have strict control on poachers and that elephant numbers are in balance in their countries. In a move to get around the ban, Botswana, Malawi, Mozambique, Zambia, and Zimbabwe set up an ivory-trading cartel that will auction tusks taken from elephants that died naturally or that were culled to keep herds at sustainable levels. But the usual buyers of ivory have agreed to the CITES decision, and are therefore unable to purchase ivory, so it is unclear who the cartel's customers will be.

A proposal on a different conservation issue has also proved controversial in international circles. Some environmental groups and a few governments are pushing to set aside a large portion of Antarctica as a reserve, protecting a unique ecosystem forever from damage from human activities. Aside from being an important symbol as the only non-militarized, non-nuclear continent, Antarctica holds productive fisheries and plays a significant role in atmospheric and climatic balance.

Since 1959 activities there have been under the control of the Antarctic Treaty System (ATS). At the moment, eighteen industrial countries and seven developing ones are party to this treaty. Seven of these twenty-five have staked out formal claims covering 85 per cent of the territory, but even the other members of ATS do not recognize these claims as legitimate. The negotiation process on further development is therefore difficult, to put it mildly.

The environmental community around the world is pressing for developing countries to have more say in this whole process, so that the future of this fragile ecosystem can be managed for the benefit of all humanity. Recent attempts to enact a Convention on the Regulation of Antarctic Mineral Resources Activities, an indefinite ban on minerals development, were thrown into disarray when the Australian government changed its mind and decided not to sign. France then joined Australia in pushing for a new comprehensive convention on environmental protection. Although this could cover the issue of minerals development, for the time being the fate of conservation in Antarctica is in limbo.

Protection of another ecosystem, and of some of the species in it, became a little more likely in late 1989 when a UN General Assembly committee approved a phased-in ban on drift-net fishing. These huge nylon nets used on commercial fishing boats can stretch forty miles across the ocean, scooping up and killing far more marine life than needed, or than can be handled. The practice is to cease in the South Pacific, where boats from Japan and Taiwan are much in evidence, by July 1991, and to halt elsewhere a year later. Although this UN resolution is not legally binding, wide observance of it is expected, given the international public outcry about drift nets.

The full range of biological-diversity issues is being addressed in a major report now being prepared by the World Conservation Union (formally, the International Union for Conservation of Nature and Natural Resources, or IUCN), UNEP, and the World Wide Fund for Nature. The *World Conservation Strategy for the 1990s* (the study's working title) applies lessons the three partners have learned since they collaborated on the first strategy, in 1980. The constituency this unique group can call on to develop the new strategy makes it by definition a global response to global problems, for those commenting on and contributing to the report represent governments, scientific bodies, and the major environmental groups in the world as well as hundreds of smaller groups.

Directions and targets will be set out in the new strategy, with an emphasis on issues that have the widest effect on other key problems and that are most amenable to action. This report, along with a broad strategy and an action plan on biodiversity being developed by WRI, IUCN, and UNEP, is bound to give a boost to the push for a global convention on biodiversity. IUCN and UNEP have been working on a draft convention for some time, in the belief that it would provide a permanent international forum for establishing priorities for conservation and investment.

The convention on biodiversity could join the treaty on the atmosphere as a concrete outcome of the 1992 Conference on Environment and Development. It could do for biodiversity what the Vienna Convention did for the ozone layer: allow governments to acknowledge the existence of a global problem, and pave the way for detailed protocols that tie governments down to specific targets and funding mechanisms.

*   *   *

'The tyranny of the immediate.' That phrase sums up well what politicians who face voters, business managers who face bottom lines, and people in the Third World who face a desperate struggle to survive come up against day after day. Immediate needs, short-term interests: these are powerful obstacles to sustainable development.

The phrase turned up in a statement from the Oslo Conference on Sustainable Development, a meeting in July 1988 called by Prime Minister Brundtland as Chairman of the World Commission. The Secretary-General of the United Nations, the heads of twenty-two UN organizations, and the leaders of UN regional economic commissions and allied financial institutions met to explore how the UN system could promote sustainable development. To get away from the tyranny of the immediate, 'a new global ethic is needed', they said, 'based on equity, accountability and human solidarity—solidarity with present and future generations'.

The participants in Oslo 'reaffirmed that sustainable development is a common objective of the UN system, including the financial institutions'. Similar support for the Commission's main messages had been heard in New York a few months earlier. The UN General Assembly passed a resolution in late 1987 welcoming the report and agreeing that it is imperative to influence the sources of environmental problems—human activities—in order to provide for sustainable development.

The resolution also called for follow-up conferences at the national, regional, and global level to be held, echoing the 'Call for Action' that closes *Our Common Future*. The Commission had noted that 'we have been careful to base our recommendations on the realities of present institutions, on what can and must be accomplished today. But to keep options open for future generations, the present generation must begin now, and begin together, nationally and internationally.' Follow-up conferences held in Kampala and Bergen and scheduled for Santiago and Bangkok aid in that process, and serve as a bridge to the 1992 Conference on Environment and Development.

Two major organizations within the United Nations deal with components of sustainable development: the UN Environment Programme and the UN Development Programme (UNDP). As *Our Common Future* put the final stamp of approval on the understanding that it is foolish to try to deal with environment and development separately, some might argue that it would make sense for these two UN organizations to combine into a Sustainable Development Programme. The size of these well-established bureaucracies and the understandable loyalties within them make such a change highly unlikely, but the 1992 Conference might come up with some overarching secretariat that works to eliminate redundancies and maximize the contributions of each. (Chapter 7 surveys various proposals for better global management.)

UNEP arose out of the 1972 UN Conference on the Human Environment in Stockholm. In the first decade of UNEP's existence, the world paid more attention to oil-price increases than to disappearing forests and the link between poverty and natural-resource degradation. Yet during this time UNEP established computerized networks for collection and monitoring of environmental information, including details of potentially toxic chemicals; set up a Regional Seas Programme that was particularly successful in the Mediterranean; helped implement the convention on endangered species; and began the difficult task of negotiating a treaty on protecting the ozone layer. In 1987 UNEP submitted to the General Assembly its *Environmental Perspective to the Year 2000 and Beyond*, a document prepared by the governments in UNEP's Governing Council.

The most recent UNEP Governing Council selected six areas for priority attention: climate change and ozone depletion, shared freshwater resources, regional seas and their coastal areas, desertification and deforestation, conservation of biological diversity, and the management of hazardous wastes and toxic chemicals. As indicated throughout this chapter, UNEP has played a key role in producing treaties in several of these areas and in working towards agreements in the others.

UNEP Executive Director Tolba is determined that 'the decade of the environment should also be the decade of decision'. He told the 1989 UN General Assembly, 'the environmental crisis demands nothing less than a revolution in the conduct of international affairs—one that acknowledges the need for global partnership and new sources of finance that will help us all, and in particular the developing nations, overcome the environmental destruction that menaces the peace and stability of the international community'.

UNDP, founded in 1951, is more involved with the design and funding of development projects in the Third World, through its 112 field offices and current support of 5,300 projects valued at $7.5 billion. By the end of 1988 more than 400 projects targeted directly at environmental issues were being funded, totalling over $300 million. The organization has an environmental action plan for the nineties that focuses on expanding the transfer of technology to developing countries and strengthening such nations' institutional and managerial capacity to formulate and implement sustainable development programmes.

One contribution to that strengthening is likely to be a new Sustainable Development Network. UNDP announced in January 1990 that it hopes to find funds to establish links among centres within developing countries that are doing relevant work in related fields. The network would also encourage and support the setting up of national committees for sustainable development, with representatives from government, industry, research institutes, and non-governmental organizations, somewhat akin to the Round Tables in Canada (see Chapter 7).

The central focus of UNDP's work remains the eradication of poverty, for as Administrator William Draper told the General Assembly in 1989, 'in every part of the globe, it is poverty and the poverty of opportunities which cause human suffering, environmental degradation and ecological depletion. Equitable economic and social growth in the developing world is a necessary condition for preservation of the environment.'

The other principal source of funds for equitable economic and social growth is the multilateral development banks—the World Bank and the regional financial institutions for Africa, Asia, and Latin America. The Brundtland Commission noted that the World Bank had:

> taken a significant lead in reorienting its lending programmes to a much higher sensitivity to environmental concerns and to support for sustainable development. This is a promising beginning. But it will not be enough unless and until it is accompanied by a fundamental commitment to sustainable development by the World Bank, and by the transformation of its internal structure and processes so as to ensure its capacity to carry this out.

This fundamental commitment was signalled, the Bank claims, by a major overhaul of its Environment Department in spring 1987. Over the next two years, professional staff working on environmental issues increased sevenfold. By fiscal year 1989, the World Bank claimed that more than one third of the projects it approved contained significant environmental components. In September, an 'environmental assessment' process that had been evolving piecemeal in the regions was formalized. These reports are intended to alert project designers, borrowers, and the Bank itself to potential problems.

In a letter with the Bank's comments on the signs of hope for our common future, World Bank Environment Department Director Ken Piddington wrote:

> Sustainable development has now become a household term. In effect, the Brundtland Commission's report has perhaps had its greatest impact in the realm of consciousness-raising and as a catalyst to further discussions on the environmental agenda that it defined so well. Particularly welcome is the report's emphasis on the need for a shift away from purely technical solutions to a broader strategy that includes, along with technology, increased local participation, changed economic policies, altered institutional structures and modified legislative approaches.

That the Bank's chief environmentalist embraces the imperative of sustainable development is encouraging, but he went on to acknowledge that 'true integration of environment and development is a challenge of exceptional proportions requiring continual self-evaluation'. Changing a bureaucracy this big is a bit like altering the direction of an ocean liner: once it starts turning, it takes a maddeningly long time for it to be fully headed in the right direction. A late 1989 progress report from the Bank admitted that, 'despite the effort made in preparing environmental issues papers and environmental action plans, firmer links still need to be established with economic and sector work'. In translation, this means that the Bank economists who deal day in and day out with their counterparts in developing-country ministries of agriculture, energy, transport, industry, and even the treasury do not yet stop to consider the environmental impact of proposed policies.

The Bank's recent move to introduce environmental-impact criteria that developing countries must meet before loans are approved will only work if the additional funds it also asked for are provided. Developing-country governments pressed for cash and with fledgling environmental departments cannot easily jump through new 'green' hoops. And they resent being told to. The Third World's concern about funding became acute in winter 1989/90 when industrial countries and the World Bank rushed to lend a hand, and a few dollars, to newly democratic East European governments. This unanticipated draw on resources for development—which comes from an allegedly finite pot of money—makes the Bank's plea for additional resources all the more pressing.

Distrust about the conditions attached to loans ('conditionality') is part of the reason the Brundtland Commission called for a substantial increase in resources available to the World Bank and the International Development Association. One response has been a new Technical Assistance Grant Program at the World Bank, established in 1989 with an initial pledge of $20 million from the Japanese government. Borrowers can apply for outright grants that will help them strengthen local institutions and their ability to prepare environmental assessments.

The institutions that provide funds for development in different regions of the world are similarly working towards environmental assessment procedures. The Inter-American Development Bank (IDB), for example, which funds projects throughout Latin America, decided in 1989 to collect the natural resources offices within its various divisions into one overall Environmental Protection Division. But this does not mean the issues are only of concern to this division. IDB President Enrique Iglesias told a May 1989 conference that 'environmental awareness must permeate our entire organization. . . . Of course, we must have environment specialists, but we must begin a process of consciousness raising for all personnel.'

To aid all these offices, as well as bilateral development assistance agencies, the Organisation for Economic Co-operation and Development (OECD) recommended that all its members consult a checklist it issued in March 1989. Asserting that a firm alliance must be forged between economic growth and environmental management, this thirteen-point summary aims to help decision-makers consider the environmental impact, mitigation measures, procedures, and implementation process of a project before approving or disapproving it.

OECD's list asks, for example, what lessons from previous similar projects have been incorporated into the environmental assessment of the current one. Another question, one of the most important, is: 'Have concerned populations and groups been involved and have their interests been adequately taken into consideration in project preparations?' Earlier failures to do so have doomed some development schemes that might have looked good on paper but that ran into human walls of tree huggers or rubber tappers out in the real world. Sustainable development will not exist until all development-assistance proposals can answer 'yes' to this key question.

* * *

Numerous international bodies that do not fall into any of the neat categories mentioned so far have discussed, debated, and endorsed the Commission's call for sustainable development. Among these intergovernmental groups, as they are called, are the Organization of African Unity, the Inter-Parliamentary Union, the Commission of the European Communities, and the Association of South-East Asian Nations.

One particularly important group to comment on the growing threats to our common future was the Heads of State or Government of Non-Aligned Countries. The final statement from their 1989 summit meeting included a statement on the environment (see Box 2–2) and a call for the accelerated economic and social development of developing countries that is a prerequisite for sustainable development.

The following month, many of these same leaders of the Third World travelled to Malaysia for the biannual meeting of the

Box 2–2.   The Summit of Non-Aligned Nations

The Declaration issued at the close of the Ninth Conference of Heads of State or Government of Non-Aligned Countries, in September 1989, included several paragraphs on environment and development:

- The world is at a crossroads: tension is no longer at breaking point but neither is peace stable; stagnation is not general but neither is development. While there may be reason for hope, there is no cause for undue optimism. The world must decide which way to turn, as we face new challenges as well as new opportunities. . . .

- There has been a ferment of new economic and political ideas in many parts of the world. These provide a propitious setting for fresh initiatives. On the other hand, if economic imperatives, and particularly the requirements of developing countries, are not accommodated, the resulting strains may very well undermine the current trends towards global peace and harmony. A detente devoid of economic content is unlikely to endure. . . .

- The somber contrast between enormous military expenditure and dire poverty underlines the importance of giving concrete shape to the concept of the link between disarmament and development. Given enhanced disarmament prospects, new opportunities are opening for all countries, especially those possessing the largest nuclear and conventional arsenals, for rechannelling additional financial resources, human energy and creativity into development. . . .

- As we approach the 21st century, protection of the environment has emerged as a major global concern, dramatically emphasizing the growing interdependence of the world. This calls for urgent co-operative measures and global compact ensuring a sustainable and environmentally sound development. Such cooperation should take place within the overall framework of the objective of reviving growth, creating a healthy, clean and sound environment and meeting the basic needs of all. Multilateral approaches need to emphasize supportive measures, while seeking to redress existing asymmetries. The international community must set aside net additional financial resources for environmental cooperation and facilitate developing countries' access to environmentally safe technologies. . . .

- There exist broad possibilities for joint action aimed at protecting and promoting the environment at the level of the entire international community within the context of a comprehensive developmental effort. We shall make our contribution to that end.

Source: Excerpted from 'Declaration', Ninth Conference of Heads of State or Government of Non-Aligned Countries, Belgrade, 7 September 1989.

Box 2–3. The Langkawi Declaration on Environment

The Heads of Government of the Commonwealth introduced the programme of action they endorsed in October 1989 with the following paragraphs, which summarize their views on the environment:

■ We, the Heads of Government of the Commonwealth, representing a quarter of the world's population and a broad cross-section of global interests, are deeply concerned at the serious deterioration in the environment and the threat this poses to the well-being of present and future generations. Any delay in taking action to halt this progressive deterioration will result in permanent and irreversible damage.

■ The current threat to the environment, which is a common concern of all mankind, stems essentially from past neglect in managing the natural environment and resources. The environment has been degraded by decades of industrial and other forms of pollution, including unsafe disposal of toxic wastes, the burning of fossil fuels, nuclear testing and non-sustainable practices in agriculture, fishery and forestry.

■ The main environmental problems facing the world are the 'green-house effect' (which may lead to severe climatic changes that could induce floods, droughts and rising sea levels), the depletion of the ozone layer, acid rain, marine pollution, land degradation and the extinction of numerous animal and plant species. Some developing countries also face distinct environmental problems arising from poverty and population pressure. In addition, some islands and low-lying areas of other countries, are threatened by the prospect of rising sea level. (cont. on p. 36)

---

forty-nine Heads of Government of the Commonwealth. This is an important forum for the Presidents and Prime Ministers who represent one quarter of the people in the world—the leaders of Australia, Canada, New Zealand, and the United Kingdom sit down with those from Bangladesh, Ghana, India, Nigeria, Pakistan, Zambia, and Zimbabwe, among others.

The usual battles about apartheid in South Africa occurred. But in addition, according to Commonwealth Secretary-General Ramphal, 'environment loomed very large indeed' at the meeting. The Heads of Government issued the Langkawi Declaration (see Box 2–3), which committed them to a sixteen-point programme of action on specific matters such as energy conservation, the need to develop international funding mechanisms, a convention on the global climate, and sustainable forest management.

Box 2–3. *(continued)*

■ Many environmental problems transcend national boundaries and interest, necessitating a co-ordinated global effort. This is particularly true in areas outside national jurisdiction, and where there is transboundary pollution on land and in the oceans, atmosphere and outer space.

■ The need to protect the environment should be viewed in a balanced perspective and due emphasis be accorded to promoting economic growth and sustainable development, including eradication of poverty, meeting basic needs, and enhancing the quality of life. The responsibility for ensuring a better environment should be equitably shared and the ability of developing countries to respond be taken into account.

■ To achieve sustainable development, economic growth is a compelling necessity. Sustainable development implies the incorporation of environmental concerns into economic planning and policies. Environmental concerns should not be used to introduce a new form of conditionality in aid and development financing, nor as a pretext for creating unjustified barriers to trade.

■ The success of global and national environmental programmes requires mutually reinforcing strategies and the participation and commitment of all levels of society—government, individuals and organisations, industry and the scientific community.

Source: Excerpted from 'The Langkawi Declaration on Environment', Commonwealth Heads of Government Meeting, Kuala Lumpur, Malaysia, 21 October 1989.

The Langkawi Declaration's call for 'a balanced perspective' in the need to protect the environment provides an important counterweight to the Group of Seven's statement of three months earlier. In Paris, all the talk about the environment focused on the symptoms of a planet in distress, and on their sources in consumption patterns in the industrial nations. The final communiqué did include in separate sections important paragraphs on general problems of development and the need for a strengthened debt strategy. But, as noted earlier, the Group of Seven Summit failed to draw the link between resource depletion and rising poverty.

Former Commissioner Ramphal explained yet again the folly of this at the Tokyo Conference in 1989:

I believe we have to take a fresh, and urgent, look at the link between environment and development and at how it can properly be addressed in international relations. *Our Common Future* brought several major insights to bear on this relationship; but some of them are being overlooked or ignored. Central among them was the conviction that mass poverty in the world is not merely unacceptable and unnecessary but that, in environmental terms, it is both exacerbated by and contributes to environmental stress. Ecology and economy, we emphasised, were inseparable. Since then, lip service has been given to the proposition; but it is clear that it is still not widely understood.

Transforming lip-service into meaningful efforts to end poverty is the challenge that global institutions and world leaders face in the nineties.

3

The Wheels of Government

> The condition of the environment is extremely serious. In spite of improvements in certain areas, the situation as a whole is continuing to deteriorate. It would be irresponsible to delay drastic measures any longer. Radical decisions, which will affect everyone, are unavoidable. Not only the improvement of environmental quality, but also the very survival of mankind is at issue. Unless we set a different course quickly and resolutely, we are heading for an environmental catastrophe. The only way to avoid it is to lay a basis now for sustainable development.

A call to action from Greenpeace? The start of the last, desperate annual report from Friends of the Earth? So it would seem. Actually, this is the opening paragraph of a publication from the Government of the Netherlands, highlighting its May 1989 national environmental policy plan, which was based on the findings of the World Commission on Environment and Development. It contains some of the strongest language seen in an official document on the environment. But the Dutch are certainly not alone in being alarmed, nor in finding that the Brundtland Report helped them to rethink their policies.

The Dutch report is one of several official responses to *Our Common Future* described in this chapter, along with some recent noteworthy national policies. Steps taken by governments at opposite ends of the earth—in Norway and Australia—are detailed in boxes to illustrate the range of issues touched by sustainable development.

The second part of the chapter looks at initiatives on specific components of our common future—clean air, a stabilized climate, the sustainable use of the earth's biological resources, an easing of the Third World's burden of debt. The condition of the atmosphere, both in terms of global warming and a depleted ozone layer, has received the most attention over the last three years, but some governments have also taken steps to halt soil erosion and the destruction of rain forests.

* * *

One of the few governments with little to say so far about sustainable development *per se* has been that of the United States. Although George Bush ran on a ticket of being the environmental President, in his first year in office he took no bold steps to live up to that pledge. The administration concentrated on one area needing immediate attention—the quality of the air—and had to focus considerable attention on damage control off the coast of Alaska. But no attempt was made to look at the bigger picture of sustainable development, with all that entails for policy changes needed across the board.

In January 1990 President Bush did announce one change that will improve the standing of the environment in the US government. He agreed with congressional advocates that the time had come to give the Environmental Protection Agency a place at the most exclusive table in town—the Cabinet. Acknowledging that many other countries already have environment ministers, Mr Bush said he hoped having a US Secretary of the Environment 'will help influence the world's environmental policy'.

Many other governments have taken the Commission's report as a starting-point for a new way of thinking about the environment. Margaret Thatcher appears to have undergone a transformation from Iron Lady to Green Goddess, as *The Economist* put it. In a letter to the Centre for Our Common Future about the impact of the Brundtland Report, Prime Minister Thatcher wrote: 'I was pleased that Mrs. Brundtland chose London as the venue for launching *Our Common Future* in 1987. We took advantage of this opportunity to welcome the report's clear analysis of the prescription of the concept of sustainable development.'

The United Kingdom was one of the first to issue an official response to the Commission. In July 1988 it published *Our Common Future: A Perspective by the United Kingdom*, a point-by-point discussion of the report and where the UK stands on the many issues it addressed.

The Prime Minister's subsequent public endorsement of sustainable development is often cited as an indication of the changed political climate surrounding environmental issues. By November 1989, in a major speech at the United Nations devoted entirely to the environment, Mrs Thatcher warned:

> While the conventional, political dangers—the threat of global annihilation, the fact of regional war—appear to be receding, we have all recently become aware of another insidious danger. . . . It is the prospect of irretrievable damage to the atmosphere, to the oceans, to earth itself. . . .
>
> Put in its bluntest form: the main threat to our environment is more and more people, and their activities: the land they cultivate ever more intensively, the forests they cut down and burn, the mountainsides they lay bare, the fossil fuels they burn, the rivers and seas they pollute. The result is that the change in future is likely to be more fundamental and more widespread than anything we have known. . . .
>
> That prospect is a new factor in human affairs. It is comparable in its implications to the discovery of how to split the atom. Indeed, its results could be even more far-reaching.

In September 1989 the Thatcher government released *Sustaining Our Common Future: A Progress Report by the United Kingdom on Implementing Sustainable Development*. And in December, as part of the effort to implement sustainable development, the government introduced an Environmental Protection Bill that proposed an integrated approach to control of air, land, and water pollution, plus stricter standards on waste disposal.

The other official responses to *Our Common Future* have for the most part been issued by West European governments. Norway has published two reports (see Box 3–1). Sweden also issued a response, and then enacted a comprehensive 'Environment Policy for the 1990s' in the summer of 1988. An Environmental Advisory Council has been appointed within the Swedish Prime Minister's office with not only representatives of government ministries but also scientists and members of outside environmental groups. A Finnish Commission for Environment and Development reported to the government in March 1989 on the full range of sustainable-development issues.

In Denmark, an action plan for environment and development was published in December 1988. It included a recommendation for a general education campaign on the many issues raised in the Brundtland Report. By mid-1989 this 'Danish Campaign for Our

Box 3–1. Norwegian Policy Regarding Global Sustainable Development

In June 1988 the Minister of Foreign Affairs in Norway published *Environment and Development: Norwegian Policy regarding Global Sustainable Development* as an initial response to the World Commission report. In acknowledgement of the cross-cutting nature of sustainable-development issues, this small booklet contributed to Norway's effort to co-ordinate the work and policies of various ministries responsible for following up the report.

By April 1989 a more complete report was ready and presented as a White Paper to the Storting (Parliament), with chapters on Norway's environmental situation, international co-operation, national goals and measures, and economic policies for sustainable development. Regarding the impact of *Our Common Future*, the government noted that 'two years after the report was presented, we can ascertain that there is new life in international co-operation on environmental issues'.

Among the many specific national goals and international policy initiatives in the 1989 report are:

■ a commitment to work towards the establishment of an international climate fund within the UN and, if other industrial nations support the idea, to contribute 0.1 per cent of Norway's gross national product to the fund.

■ a stabilization of carbon dioxide emissions by the year 2000 at the latest, which when combined with goals to reduce CFC and nitrogen oxide (NO_x) emissions would lead to an overall reduction of greenhouse gases; although Norway contributes only 0.2 per cent of global CO_2 emissions, this commitment was the first by a government to slow global warming, with the understanding that 'our national efforts will primarily be meaningful when they help to hasten a broad-based process involving several countries'.

■ a levelling-off of total energy consumption around the end of the nineties, in part through the use of pricing and taxation policies that reflect the environmental costs of various energy sources, and a move towards forms of energy that pollute less.

Common Future' was launched, with an overall theme of 'Think Globally—Act Locally'. Two special themes picked for 1990 are Our Common Air and Our Common Consumption. A total of 21 million kroner ($3.25 million) has been appropriated by the Folketing (Parliament) to support national and local activities that promote new forms of co-operation.

The campaign office in Copenhagen stresses that it is action that will be funded, not more reports and more pieces of paper. Hugo

- a 50-per-cent reduction of CFC emissions in 1991 and at least a 90-per-cent drop by 1995 (both based on 1986 emission levels), with a ban on the manufacture or import of products containing CFCs after 1995.

- the halving of sulphur dioxide emissions from their 1980 levels by 1993, and NO_x emissions by 1998 that are 30 per cent below 1986 levels.

- discharges of nutrient salts to vulnerable areas of the marine environment by 1995 that are half the levels of 1985, in line with 1987 agreement of the North Sea states, and, going further than that agreement, discharges of toxic substances by then that are reduced by around 70 per cent.

- a reduction of pollution from hazardous wastes by the year 2000 to the point where they pose no danger to human health or the environment; the establishment of a central treatment plant in Rana and, by 1991, of 200 delivery sites for hazardous wastes; and a ban on the export of hazardous wastes to developing countries.

- a call for greater weight for environmental issues in the work of the World Bank, which will be aided by new instructions being prepared for the Nordic director at the Bank.

- a proposal to establish within the United Nations an ecological security council, suggested as 'a symbol and an example of a more binding form of international co-operation on global environmental issues'.

- a status report to be delivered to the Storting in 1990 on national and international progress on the World Commission's recommendations.

Source: Ministry of Environment, *Environment and Development: Programme for Norway's Follow-up of the Report of the World Commission on Environment and Development*, Report to the Storting No. 46 (1988–89) (Oslo, 1989).

Prestegaard of that office sees those involved as 'ambassadors for action'. They must forge new partnerships among groups that do not normally listen to each other, forcing a dialogue on the changes needed. In nine local areas, this will be aided by 'green councils' that the government has encouraged (see also Chapter 7). The government plans to build, it says, on the strong Danish tradition of active environmental work: 'The task is not to make people know. The task is to make them act. Only action creates change.'

As the startling language in the beginning of this chapter indicates, *Our Common Future* was taken to heart in the Netherlands. Indeed, in an October 1989 speech to the International Environment Forum in New York, Minister Ed Nijpels noted that the Brundtland Report 'has made a major contribution to a different view of the relationship between environment and economic development'. Under the current plan, by the end of this decade the Dutch will be spending 5 to 6 per cent of their national income on the environment—nearly twice the amount they spend on defence.

A telling sign that the Dutch took seriously the challenge to change is that the national environmental-policy plan was drafted jointly by four ministries: Housing, Physical Planning and Environment; Agriculture and Fisheries; Transportation and Public Works; and Economic Affairs. As *Our Common Future* pointed out, 'the major central economic and sectoral agencies of governments should now be made directly responsible and fully accountable for ensuring that their policies, programmes, and budgets support development that is ecologically as well as economically sustainable'.

Having ministries co-operate to produce plans like this is one good way to ensure that this happens. Perhaps the most important legacy of *Our Common Future* will be the pervasive change in environment's place in the machinery of government, a change illustrated by the way the Dutch policy plan was developed. Concern for the environment has become an integrating factor.

The goal of the policy plan from the Netherlands is to solve environmental problems within one generation—twenty to twenty-five years—although the report acknowledges that for some problems this will simply be impossible. The plan is also notable for identifying consumers as a target group, one we all belong to, as it notes. By the year 2000 consumers are expected to have 50 per cent of their organic waste composted or fermented; separate their household trash into tin, glass, textiles, paper, used batteries, and small chemical wastes; and drive 15 per cent fewer kilometres than they did in 1985.

Half a world away, the Ministry of Environment in Japan published a *White Paper on the Environment in Japan 1988*. It was the first time in the twenty years of these white papers that the focus was the global environment. As in the Netherlands, the importance of educating consumers was noted: 'It is necessary to conduct environmental education putting the earth at its center and to explain in plain language the relationship of human beings with the global environment and the necessity of sustainable development.'

To oversee all government programmes in this field, a Cabinet-level Conference of Ministers for Global Environmental Conservation was set up in May 1989. As Japan is the second-largest economic power in the world, and the largest donor of development aid, the new council could play a leading role in securing our common future.

The necessity of sustainable development is equally clear to many governments in the Third World, and often more pressing as they struggle to meet the needs of a rapidly growing population with a rapidly deteriorating natural-resource base. In recent years, a number have prepared action plans or strategies, sometimes in co-operation with UN agencies, the World Bank, or the World Conservation Union. These are not in response to the Brundtland Commission, although some of them draw on *Our Common Future* when describing their intentions to introduce sustainable development throughout the government.

The National Environmental Action Plan for the Kingdom of Lesotho, a small nation surrounded by South Africa, is but one example of this new approach. After discussions with non-governmental groups, business representatives, academics, and local district legislators at special conferences and in private meetings, and after wide review of several drafts, the action plan was approved by the Cabinet in May 1989. Its opening paragraphs note that 'in Lesotho, the organic link between the environment and development is all too clear'.

A comprehensive review of the nation's problems and current government programmes is followed by detailed recommendations in eleven areas, such as how to improve livestock marketing, enhance soil fertility and crop-land productivity, broaden the availability of family planning, and use and dispose of hazardous chemicals properly. The government is working with the UN and multilateral development banks to find the funds to carry out the action plan. Many more such strategies or action plans can be expected in the nineties—the names will differ but the intent will be the same—as developing countries consider how to translate the theory of sustainable development into the reality of government programmes.

In Australia, a July 1989 announcement by Prime Minister Bob Hawke—*Our Country, Our Future*—was called the most significant statement on environmental issues in that country to date (see Box 3–2). The government's efforts in this field also include support for the Commission for the Future, which tries to raise awareness

Box 3–2. Australia's *Our Country, Our Future*

In July 1989 Prime Minister Hawke gave a speech on the environment and released the government's most complete statement on the subject to date, entitled *Our Country, Our Future*. 'The threat posed by continuing environmental deterioration is no longer hypothetical', he noted, and 'we have little time to spare.' Nearly two thirds of the land in Australia requires treatment for degradation, and forest cover has been cut in half since European settlement began. The government makes no bones about having the worst record in the world on mammal extinctions.

The Prime Minister stated clearly that the government 'is committed to the principle of ecologically sustainable development'. The report discusses Australia's role in global environmental issues and, at the national level, steps to be taken on natural ecosystems, the atmosphere, land, and humans and the environment.

Among the specific steps taken in Australia to support this national commitment are:

■ the Ozone Protection Act of 1989, which provided for a ban on the manufacture and import of polystyrene packaging, insulation material, and virtually all aerosols containing CFCs after 1989; regulations introduced in late 1989 will mean the phase-out of nearly all CFC and halon use by the end of 1994.

■ the establishment of an Environmental Resources Information Network that will collect and supplement information on endangered species, vegetation types, and heritage sites.

■ the establishment, in 1988, of a Resource Assessment Commission as an independent body to deal with especially complex and contentious resource issues; it will first assess forests and timber resources and then consider coastal-zone resources.

■ the adoption of new guide-lines to improve integration of environmental protection and resource management into Australia's development-assistance programme, pegged at Aus.$1.1 billion in 1988–89, and the establishment of a special four-year Environment Assistance Programme worth Aus.$20 million.

of long-term issues facing Australians. The Commission's adaptation of a US group's *Personal Action Guide for the Earth* (see Chapter 6) was widely distributed and even reached an estimated two million Australians over the radio.

Nearby, the government of New Zealand introduced a Resource Management Bill in late 1989 that it called a 'framework rather than a blueprint'. It pulled together all previous laws on land,

- the appointment of an Endangered Species Advisory Committee with representatives from the government, scientific institutions, the farming community, and non-governmental conservation groups who will develop a national strategy to halt the loss of species.

- a ban in Australian waters on drift-net fishing and on the shipment of any fish caught by such methods, while the government works for a global ban on drift nets.

- a decision in May 1989 not to sign the Antarctica Minerals Convention because it accepts the principle of mining, albeit under some safeguards, and to work instead on a comprehensive environmental convention for the continent, including the establishment of a wilderness reserve.

- the 1989 release of a National Soil Conservation Strategy, prepared by the new Australian Soil Conservation Council; a Ministerial Task Force on Soil Conservation has also been established, with representatives of farming and conservation groups joining those from the government.

- the declaration of 1990 as the Year of Landcare, the first in a Decade of Landcare that is due to receive more than Aus.$320 million in government funds for soil conservation, tree planting, and related vegetation conservation programmes.

- a pledge to have one billion trees planted and growing by the year 2000, in a programme that started in 1989 with government funding of Aus.$4 million; community groups, corporations, and schools are to be involved in the planting campaign.

- the appointment of Sir Ninian Stephen, former Governor-General of Australia, as the country's first Ambassador for the Environment, to advise the government and represent it in international negotiations on climate change, biological diversity, overfishing, forestry management, and Antarctica.

Source: *Our Country, Our Future*, Statement on the Environment by the Hon. R. J. L. Hawke, Prime Minister of Australia (Canberra, Australian Government Publishing Service, 1989).

water, air, and mineral resources. Under the term 'sustainable management' (which New Zealand defines just as the Commission did sustainable development), the Ministry of Environment notes: 'Integration is the key. A single set of consistent objectives for resource management is provided. . . . Clearer roles have been set down for central government and regional and territorial councils. The rights and responsibilities of industry and the wider community are firmly established.'

A similar major policy initiative on the environment in Canada— also thought of as a framework document—is expected in late

1990. This will be the next step in one of the most wide-ranging government responses to the World Commission on Environment and Development. Delays in getting agreement on this report may indicate, however, that the machinery of the Canadian government is not yet ready for an approach that integrates ecology with the economy.

Nevertheless, before the World Commission's report was even delivered to the UN General Assembly, a National Task Force on Environment and Economy appointed by the Canadian Council of Resource and Environment Ministers published a report on the implications for Canada. Environmental writer Michael Keating calls this booklet a historic statement not only for what it said but for the people who said it—corporate leaders, environmental groups, and government ministries, who joined together and agreed 'change is necessary and it must occur now'.

One of the group's forty recommendations was that ongoing round tables on environment and economy should be formed throughout the country, to act as clearing-houses for innovative solutions to mounting problems. (See Chapter 7 for more on this new way of thinking about institutions.)

Another mechanism for tracking solutions is the Success Stories Bank at Environment Canada, the federal ministry. Launched in late 1988, the office is working to identify and publicize companies and government programmes that are taking steps to improve or not degrade the environment. It is a section within the ministry's Sustainable Development Branch, which is a rather unusual title for a government office. In theory, sustainable development should apply to and pervade all government agencies. But until that happens, having staff whose specific goal is to further the concept is commendable—something like an affirmative action programme for the future.

The Canadian government's avowed aim is to become a model environmental institution. And it publishes a wealth of information (increasingly on recycled paper, it hopes) to help others do the same. It also launched an Environmental Partners Fund in 1989 to encourage local activities. Stressing that individuals can make a difference, the government pledged over five years to match Can.$50 million ($42 million) in funds raised by local groups. Projects will be supported with up to Can.$200,000 ($167,000) over a three-year partnership; they can focus either on conserving or rehabilitating the environment or on recycling or reducing waste. The programme is specifically targeted at new projects, not ones required by law or regulation.

Canadians often speak of the need to involve all the 'stake-holders' in working towards sustainable development, an appealing concept because the word is such an inclusive one. Everyone, after all, has a stake in seeing that the resources to provide food, clean water, breathable air, and roofs over our heads remain available for ourselves and for the next generation. The Canadian approach of drawing everyone in, of creating partners in the effort to secure our common future—like the Danish campaign—is exactly what the World Commission had in mind when it noted: 'The law alone cannot enforce the common interest. It principally needs community knowledge and support, which entails greater public participation in the decisions that affect the environment.'

*   *   *

These ringing endorsements of *Our Common Future* and declarations of 'green' intentions by politicians in industrial countries are all well and good. As Michael Keating points out in a recent report for Environment Canada, 'statements of intent are the necessary precursors to action'. But the head of that same agency, Minister of the Environment Lucien Bouchard, made a more telling point in his October 1989 speech to the UN General Assembly: 'While speeches and rhetoric are important, and may often be powerful and moving, their value is limited if words do not result in concrete changes to our political and economic behaviour.'

In other words, talk is cheap. As Chapter 2 documented, government representatives are good at making speeches and signing protocols—in Montreal, Basle, Paris, and so on. The remainder of this chapter looks at what the speakers and signers have done at home about specific sustainable-development concerns.

Numerous atmospheric issues have received the most attention since *Our Common Future* was published. The problem most in evidence is, no doubt, air pollution: the build-up of sulphur dioxide (SO_2), nitrogen oxides (NO_x), carbon monoxide, and other particulate matter that is the unwelcome accompaniment of industrialization and motorization.

Dirty air has long been with us, so governments have in place more detailed solutions to this environmental problem. The various national plans described earlier in this chapter have virtually all included details of emission-reduction goals. A tally of these yields a bewildering list of who will do what, and when: Norway, for example, aims to have sulphur dioxide emissions in 1993 that are half their 1980 levels, and nitrogen oxide emissions by 1998 that are 30 per cent below 1986 levels. The Netherlands hopes to have SO_2 emissions by 2000 that are 80 per cent below 1980 levels, and NO_x ones by then that are half the 1980 level. Denmark (along with eleven other West European nations) has pledged to bring NO_x emissions 30 per cent below 1987 levels by 1998.

Although these variations will affect us all, the details make fascinating reading only for those who earn a living by tracking protocols and strategies. What will be the net result for the environment? Commitments such as these 'are far better than nothing', notes Hilary French of Worldwatch Institute. 'But they are not enough. . . . The reductions envisioned are too little and too late to protect the environment adequately. Ecologists suggest that cuts on the order of 90 percent in sulfur and nitrogen oxides and 75 percent in ground-level ozone are what is really needed.'

Attempts in the United States to revise the Clean Air Act of 1970 received a shot in the arm in mid-1989 when President Bush introduced sweeping amendments. But these were watered down somewhat during committee negotiations, and as of early 1990 had not yet been enacted. In the meantime, southern California decided to forge ahead with its own plan. The South Coast Air Quality Management District, which includes smog-ridden Los Angeles, introduced a twenty-year plan that has been called draconian by some and visionary by others. The number of cars per family will be limited, car-pooling will be required, gasoline-powered lawn-mowers and barbecues needing lighter-fluid will be banned, and all pollution sources will need emission-control devices—down to the level of restaurants that charbroil their entrées.

The plan will affect every Southern Californian, and will certainly have wider impacts. Legislation in this part of the country, home to 5 per cent of Americans, has a habit of being followed by similar laws in other states, though the specific provisions often vary. Industries affected by the new regulations then join environmentalists in a push for national legislation, as happened with energy-efficiency standards for appliances in the early eighties. To companies aiming for the efficiency of mass production, one standard is much better than fifty different state requirements.

California's neighbour to the south has come up with its own plan to clean up the air. In November 1989 Mexico City introduced a 'No Driving Today' programme in the nation's capital in response to worsening air pollution. Coloured stickers on the windows of the area's three million cars and trucks determine which day each week the vehicles must be left at home. Police at key intersections pull over drivers in violation, and the stiff fine—300,000 pesos ($132)—has apparently deterred many from trying to get around the law. Similar schemes to reduce the use of cars exist in Florence, parts of central Rome, Santiago, and Budapest when the pollution is particularly bad.

As one indication of the impact in Mexico, the state oil company, Pemex, reported that gasoline consumption dropped by about 750,000 gallons a day. The government claims that pollution levels are 15 to 20 per cent lower, although some environmentalists believe that is an overstatement. But all agree that the air is noticeably cleaner. Youngsters in Mexico City are finally able to see nearby volcanoes they have only heard about before.

'These are the first concrete steps after many promises', Homero Aridjis of the environmental alliance Group of 100 told the *New York Times*. 'What is most interesting is that for the first time there has been a civic mobilization to face the problem.' The government was being pressured by some businesses to cancel the experimental programme when the trial period ended in February 1990, but others tried equally hard to persuade it to make the scheme permanent.

A different type of emission has received a great deal of attention since *Our Common Future* was published. The impact of chlorofluorocarbons (CFCs) on the ozone layer is the issue that has most captured both public attention and local politicians' votes. And it is the issue on which there has been the most concrete progress since the Brundtland Commission completed its work.

While governments wrangle over whether to strengthen the Montreal Protocol in light of new evidence of the damage so far (see Chapter 2), many of them have gone beyond it in their own legislation. Again, a confusing mix of goals and target-dates prevails. Australia's July 1989 statement staked a claim to that nation being the only one to limit export of CFCs; in addition, nearly all CFC and halon use there is to stop by 1995. That year has also been set as the deadline for CFC use by the Federal Republic of Germany, Sweden, and the Netherlands (or 'as rapidly as possible thereafter'). Norway expects to reduce use by at least 90 per cent by 1995, and to stop halon use by the mid-nineties as well. The United Kingdom aims to cut production and use by 85 per cent as soon as possible, and to fully halt both by the end of the decade, as does Canada.

As with air pollution, cities and other jurisdictions are taking matters in their own hands. In North America, twenty-four municipalities—including Los Angeles and Toronto—joined together in a Stratospheric Protection Accord. They have agreed to ban the local use of ozone-depleting substances by early 1992 unless no technically feasible alternatives exist by then, and to require the recovery and recycling of CFCs from products such as refrigerator-coolant units.

The state of Vermont has taken similar actions, though some would call them largely symbolic. Madeleine Kunin, in an apparent bid to be the Greenest Governor in the nation, has banned the sale of car air-conditioners (which use CFCs in the coolant) after 1992. Given the average summer temperature in this New England state, however, this act has been likened to abolishing the use of snow-tyres in Florida. Still, an ambitious proposed state-wide environmental plan could serve the same function as the air-pollution strategy in southern California: become a model for those who believe that time is too short to wait for strong national leadership.

Eliminating CFC use will also help with the third major atmospheric issue: global warming. Far more important for the world's efforts to minimize climate change, however, will be reductions in both the burning of fossil fuels and the clearing of forests. As *Our Common Future* pointed out, 'carbon dioxide output globally could be significantly reduced by energy efficiency measures without any reduction of the tempo of . . . growth'. To achieve this, some governments have set, or at least proposed, national goals to freeze or even lower carbon emissions.

At the moment, the proposal in the Netherlands goes the farthest. Through adjustments in energy services, the electricity sector, and transportation, the nation aims by the mid-nineties to freeze carbon dioxide (CO_2) emissions at their 1989/90 levels, and then to begin reducing them. In January 1990 a tax on fossil fuels geared to their carbon concentration per unit of energy was introduced.

The national policy plan is quite clear that over the next twenty years 'the use of motor cars will have to be reduced'. This attitude has even been endorsed by the President of Volvo in Sweden, who has said that, in city centres at least, 'public means of transportation must, to a large extent, replace the private car'. The Netherlands proposes to encourage bicycle and public-transport use and to make it easier for people to live near their workplace.

Other current national goals for the most part aim at freezing CO_2 emissions by the end of the nineties (Norway) or stopping the growth earlier, so that they stay at the 1990 levels (Sweden). Although Prime Minister Thatcher, in her letter to the Centre for Our Common Future, called global warming 'the single most dangerous threat to our environment and sustainable development', the UK government at the moment is content to study the issue further.

In the Federal Republic of Germany, a major interim report on 'Preventive Measures to Protect the Earth's Atmosphere' was published in 1988 by a Bundestag Study Commission that has both parliamentarians and scientists on it. They heard testimony from other scientists and politicians and from industry representatives, commissioned more than 100 original research reports, and held discussions with environmental and consumer groups.

Although it nearly matched the Dutch report in the strength of its rhetoric about the urgency of the problem, this report, because it was an interim one, stopped short of setting specific targets for reductions of greenhouse-gas emissions. It did note that an international convention for the protection of the atmosphere is 'an indispensable necessity'. The interim report and its recommendations were approved by the Bundestag in March 1989. A final report, with goals and target-dates, is expected in summer 1990, and could have a major impact on governments' responses to this global problem.

In the US Congress, two bills call for cuts in carbon emissions of 20 per cent by the year 2000, although they are not given much chance of passing as they now stand. But a national energy strategy due in autumn 1990 from the administration is expected to be pro-conservation and renewable fuels, which could be an important step in the right direction. In his February 1990 speech to the

Intergovernmental Panel on Climate Change, President Bush noted that a $336-million programme on efficiency and renewables is due to produce energy savings worth $30 billion during the nineties.

Most government reports described earlier in this chapter say they are 'committed to developing a national strategy on climate change', or words to that effect, and then set no concrete goals. Variations of that old joke—everybody talks about the weather but nobody does anything about it—are heard more and more these days among environmentalists. Yet Christopher Flavin of Worldwatch says that cuts in global carbon emissions of 12 per cent by the year 2000 are feasible as well as needed. They would involve annual reductions of 3 per cent in, say, the United States and the Soviet Union, and of 0.5 per cent in countries with emissions at the level of France and New Zealand.

One proposed answer to global warming—a greater reliance on nuclear power—does not have too healthy a prognosis. Even the Thatcher government has got cold feet. In November 1989 the government pulled the nuclear facilities out of its scheme to privatize the UK power industry when it realized that there would be no takers. According to Labour MP Frank Dobson, 'this spells the end of the government's commitment to an expanding programme of nuclear power'.

With no viable plans as yet for where to put nuclear wastes, and with a growing number of plants coming to the end of their useful lives, forcing governments to deal with decommissioning problems, the acceptability of nuclear power is exactly where it was when the Brundtland Commission agreed that 'the generation of nuclear power is only justifiable if there are solid solutions to the presently unsolved problems to which it gives rise'.

*   *   *

Less measurable progress has been made since 1987 on protecting and restoring the biological resources that support life on earth. Public concern about tropical rain forests, for example, has certainly sky-rocketed. But too few governments have committed enough resources to halting the destruction.

Two that have recently taken encouraging steps are Thailand and Brazil. Floods in late 1988 in Thailand brought home forcefully one of the tragic consequences of mountain-sides stripped of trees: water rushed down hills and into unprepared villages, killing 400 and leaving tens of thousands homeless. Soon after, the government extended a decade-long ban on exports of logs into the whole commercial forest industry—all logging was forbidden. At the same time, the Royal Forestry Department renewed its pledge to have forests cover 40 per cent of the country, up from the current figure of less than 20 per cent. Early reports, however, indicate that the success of these efforts may require greater involvement of local people than the schemes now provide for.

Brazil has been the focus of intense international scrutiny over the last few years because of the rapid rate at which its forests are disappearing, taking with them the richest species habitat in the world and the homes of many tribes of indigenous peoples. The protection of the Amazon is now enshrined in the nation's new constitution, which includes articles about the role of government in conserving 'the variety and integrity of Brazil's genetic wealth', and specifies that the Amazon forest and several other woodlands 'are part of the national wealth, and they shall be used, according to the law, under conditions which ensure preservation of the environment, including the use of natural resources'.

In April 1989 President Sarney announced a five-year, $100-million programme to zone some half-billion hectares of the Amazon forest basin for economic and ecological uses. In launching 'Our Nature', a national project on the environment, the President said, 'Our Nature is more than just a program. It is our patrimony, our life.'

Mr Sarney took issue with non-Brazilian estimates of the rate of deforestation in the Amazon; he claimed that the government had better estimates of this problem than outsiders did, but assured his listeners that Brazil wants to preserve the rain forest: 'No one is better aware than we are that ecological problems cannot be separated from development. We do not wish to grow at any cost. We want to grow with responsibility.'

In October 1989 the Brazilian government put a three-month freeze on new subsidies and tax breaks for cattle-ranchers, miners, and lumber companies, a freeze that was extended once. It remains to be seen whether the new government under Fernando Collor de Mello will support these steps to preserve this vital area.

At the same time that efforts are made to stop deforestation, reforestation of already-stripped areas must proceed. China and the Republic of Korea have had major tree-planting programmes in place for a number of years. By the mid-eighties, an estimated eight million hectares of land were being covered a year in new trees in China, according to Worldwatch Institute. In India, reforestation was also given a high priority by Rajiv Gandhi while he was Prime Minister; he created a National Wastelands Development Board and called for a 'people's movement for afforestation'. In Australia, as noted in the box on Prime Minister Hawke's policy statement, the government has pledged to plant one billion trees in the next ten years.

The US government announced in January 1990 that it would also plant one billion trees—but that it would do this every year— as part of an effort to slow global warming. In the budget for fiscal year 1991, which the administration claims includes $2 billion in new spending to protect the environment, President Bush pledged to have ten billion trees planted by the decade's end. Environmentalists were quick to point out that this would offset only 1 to 3 per cent of the CO_2 produced by Americans a year. Nevertheless, if this turns out to be the start of a larger campaign to cut national CO_2 emissions, it will be a useful symbolic act.

Healthy soils are also essential for our common future. Without them, efforts to produce food for the eighty-eight million people currently added to our numbers annually are handicapped. Some twenty-four billion more tonnes of soil are lost annually through erosion than are formed through natural processes, according to Lester Brown of Worldwatch. In 1986 the United States decided to do something about the worsening situation in the world's 'breadbasket'.

The US Conservation Reserve Program, one of the few government attempts to deal with soil erosion, aimed over five years to plant grass or trees on sixteen million hectares of erodible cropland, which could easily have become wasteland if farmers continued to grow food on it. By mid-1989, 80 per cent of the target had been met, but funding was unfortunately pulled away from the programme by congressional budget-cutters.

Australia acknowledged its own severe problems in the Prime Minister's July 1989 statement: 'Soil erosion over much of the continent has risen to ten times the natural geological rate.' In response, a National Soil Conservation Strategy was released in 1989; specific targets and performance criteria are due to be set within the strategy by mid-1990.

The Australian government has declared 1990 to be the Year of Landcare, the first in a decade so named as part of the effort to stem erosion. During the nineties the government plans to spend at least Aus.$320 million ($241 million) conserving soils, planting trees, and introducing other conservation programmes. The National Soil Conservation Programme was established in 1983 to provide leadership and funds to fight soil degradation. It will receive nearly Aus.$50 million ($38 million) over two years to, among things, conduct a national assessment of land degradation.

Another agricultural issue moving higher on the public's agenda is the use, and over-use, of pesticides and fertilizers. Several governments are moving to discourage this practice, out of concern about the contamination of ground-water as well as about the residues in the food supply. The National Environmental Policy Plan in the Netherlands sets an overall goal of adding no more nitrogen and phosphorus to crops, for example, than can be absorbed over the country's entire agricultural area. And by the end of this decade, the use of pesticides should be cut in half.

In the United Kingdom, nineteen areas have been designated as Environmentally Sensitive since 1987, with restrictions on the use there of fertilizers, pesticides, and herbicides. All pesticides used in the country must be approved, and a code of practice is to be introduced in 1990. The 1988 Swedish Environmental Bill called for a cut in fertilizer use by 20 per cent by the end of this decade; one incentive towards this end for farmers is that the environmental levy on these and on pesticides was doubled.

Some governments are moving beyond pesticide regulation. A 1987 law in Denmark, for instance, gives farmers financial support to develop or convert to organic farming. An Organic Farming Council was set up in July of that year to promote this agricultural method and monitor developments.

Others are promoting integrated pest management (IPM) as an alternative to pesticides, an approach the Brundtland Commission encouraged. IPM uses a balance of natural predators of pests, planting patterns, pest-resistant crop varieties, and the sparse use of chemicals to keep weeds and insects at an acceptable level—a level, that is, that causes no significant economic losses. Farmers in Brazil and China have used IPM extensively to reduce their dependence on pesticides.

The Indonesian government may have taken some of the boldest steps to support IPM. In late 1986 fifty-seven pesticides were banned and a national promotion of less-environmentally harmful pest control began. A year later, subsidies for the use of remaining pesticides on rice were cut, as the success of IPM became clear to more farmers.

Although similar support for moving away from chemicals has yet to come from the US government, the spring 1990 debate on the Farm Bill is expected to focus extensively on such a move. This follows a major study by the National Academy of Sciences that found American farmers who use few or no chemicals are as productive as those dependent on pesticides and chemical fertilizers. The US government may be the next to decide to help farmers get off the 'pesticide treadmill'.

*   *   *

This review has covered principally government efforts to restore the environment. Yet those efforts will be wasted if governments ignore the other component of sustainable development, defined by the Commission as 'meeting the needs of the present without compromising the ability of future generations to meet their own needs'. Restoring, conserving, protecting natural resources: these are all essential if future generations are to have a chance to meet their needs. But if more is not done to meet the needs of those here today—the majority of humanity, pushed into over-farming, over-grazing, over-cutting, even over-populating, as they look to their children for extra hands to work and some comfort in the future—if more is not done for these people, the earth will be no better off. The moral imperative of aiding Third World development has been strengthened by our understanding of the links between poverty and environmental degradation.

Addressing this development issue, the Commission's recommendations on the flow of financial resources to the Third World dealt with both the quantity and the quality of aid: 'The need for more resources cannot be evaded. The idea that developing countries would do better to live within their limited means is a cruel illusion. Global poverty cannot be reduced by the governments of poor countries acting alone. At the same time, more aid and other forms of finance, while necessary, are not sufficient. Projects and programmes must be designed for sustainable development.'

For several years, Norway has been the largest donor of

development aid in relation to its gross domestic product. In 1986 that figure was 1.2 per cent; by comparison, the figures for Sweden, Canada, and the United States, to pick just three, were 0.87, 0.5, and 0.2 per cent. In contrast to the practices of most donor governments, none of the funds from Norway are 'tied'—dependent, that is, on the recipient buying Norwegian goods or services. And between 1986 and 1989 the allocations for environmental projects more than tripled. Environmental impact assessments will soon be required of all projects. (See Chapter 2 for information on multilateral banks' use of these.)

Canada also now requires such assessments. Since mid-1986 all new bilateral-aid projects of the Canadian International Development Agency have been screened for their environmental impact. New guide-lines in the Australian International Development Assistance Bureau call for procedures to perform similar assessments. In his July 1989 statement Prime Minister Hawke announced the establishment of a four-year special Environment Assistance Programme worth Aus.$20 million ($15 million).

Aid is unfortunately more than offset by debt. In 1989 developing countries received $92 billion in official development assistance; they paid out $142 billion servicing their debts, which totalled $1,165 billion at the end of that year. In other words, the Third World gave to the First $50 billion more than it received. (See also Chapter 8 regarding the debt issue.) In acknowledgement that some of these financial obligations will just never be met, a few governments have unilaterally forgiven the debt of nations in the direst economic circumstances. This approach is more applicable for Africa, where some four fifths of the debt is publicly held, than for Latin America.

In late 1987 Canada announced it was converting Can.$670 million ($561 million) worth of outstanding loans to sub-Saharan governments into grants. Britain has cancelled £1 billion ($1.7 billion) of old aid loans to the poorest nations. In July 1989 the United States announced it would do the same for $735 million it was owed by twelve sub-Saharan countries. And West Germany has told twenty-eight nations in Africa they never need to repay DM6.4 billion ($3.8 billion) of debt.

*   *   *

As this chapter has documented, the pace of change by governments is uneven. One unusual sign of hope may be a nationalistic race to claim being Number One at something. In this case, it has a rather odd ring to it: Canadian Minister of the Environment Bouchard announced that his country has the worst per capita record for energy efficiency, while Prime Minister Mulroney recently said that Canada produces the most municipal solid waste per person. Ed Nijpels, Minister of Housing, Physical Planning and Environment, claimed that the Netherlands uses the highest amount of agricultural pesticides in the world. Australian Prime Minister Hawke pointed out that his nation has the world's worst record for mammal extinction.

Thus, governments are rushing to say—yes, we know how bad we are, and we have seen the light. With the rise of green parties in France, Italy, Sweden, the United Kingdom, West Germany, parts of the Soviet Union, and throughout rapidly changing Eastern Europe, more and more legislatures are likely to be enlightened.

The next rush must be to take the strongest action against the many problems addressed in *Our Common Future*, instead of just acknowledging them. For CFC bans, the race is nearly over. For carbon dioxide emissions, it has barely begun. In April 1989 the Norwegian government staked a claim to being the first to set a goal of freezing national CO_2 output during the nineties. By September, the newly elected Dutch government went further, pledging to freeze emissions by mid-decade, and introducing the first carbon tax in the world.

If this is a new game of one-upmanship, we could all be the winners. The 1992 UN Conference in Brazil (see Chapter 8) provides a grand opportunity for nations to join in. As that event looms closer, governments may scramble to introduce concrete and significant legislation at home, as a sign of their willingness to 'save the planet'. Just as the speed of change in Eastern Europe left most of the world gasping for breath, the pace of change on the environment may be surprising once the floodgates are open.

The parallels with the breathtaking autumn of political change apply to more than just its pace. As *New York Times* columnist Flora Lewis writes: 'Much has been said about the speed of collapse in Eastern Europe. It is true that, like an avalanche, the televisual action came all of a sudden. But it is also true that, like such natural upheavals, the dynamics that triggered the movement were building up for a long time.'

So, too, the dynamics of the push towards sustainable development have been building. Perhaps, with hindsight, the celebrations and demonstrations around the world for Earth Day 1990 will turn out to have been the 'Berlin Wall' of the world's effort to secure our common future. Or maybe the floodgates will open as governments prepare for the 1992 conference. In either case, the condition of the earth is certain to be a top political concern in the nineties. As noted earlier, however, unless the condition of the people living on it also ranks high on governments' agendas, what appears to be progress towards sustainable development will turn out to have been an illusion.

4

Private Groups: A Force for Change

In October 1987 the report of the World Commission was presented to the United Nations in New York by its Chairman. The day began with Gro Harlem Brundtland's speech to the UN General Assembly, the focus of many people's hopes for truly global responses to global problems. She spoke to the gathering not only as Chairman but also as the leader of a government trying to apply the Commission's recommendations at the national level. Her final speech of the day was to several hundred people from US non-governmental organizations (NGOs) gathered for the James Marshall Memorial Lecture, an annual event arranged by the Natural Resources Defense Council. Thus in the space of one day, the work of the Commission touched the international community, governments, and private groups—three of the key sectors of society that must work together if sustainable development is to be achieved.

Many of those in the audience at the Marshall Lecture were leaders of groups working on environment and development issues. Mrs Brundtland talked about the vital role of such groups, and challenged those gathered to make sure the international diplomats she spoke to earlier as well as politicians like herself did more than just pay lip-service to the imperative of sustainable development (see Box 4–1).

This sector of society, more than any other discussed in *Signs of Hope*, must be short-changed in this attempt to provide an overview of recent developments. Yet the growing number, new roles, and expanding influence of non-governmental organizations over the last two decades is one of the most striking signs of hope for our common future.

Box 4–1.　Gro Harlem Brundtland's Challenge to Non-Governmental Organizations

The Commission found that a major prerequisite to sustainable development is a political system that secures effective citizen participation in decision-making. The NGOs have shown what an effective force they can be in bringing the realities and the concerns of local people to the attention of national governments. This happens not only in the North where communications and affluence can ease the process. It happens to an increasing extent in the developing world as well.

A major theme of *Our Common Future* is that a new multilateralism will be crucial for progress. NGOs, especially those that operate internationally, can be and are a force for change in this respect. The NGOs have been effectively informing the government's decisions for the northern industrialized nations for decades. . . . In the developing nations, a younger NGO movement has been increasingly effective in opening and deepening channels of communication between governments and their citizens. . . .

Up until a few years ago, the fragmented concerns of NGOs all too accurately mirrored the too fragmented concerns of the governments and their institutions. There were conservation groups, development groups, relief groups, women's rights groups, population groups, and disarmament groups. Too often they competed against one another more often than they cooperated; too seldom did they seek out common ground.

It is also one of the hardest to pin down. There are about as many definitions of NGOs as there are of sustainable development. Earnest researchers fill journals with articles about the differences between a private voluntary organization, a service organization, and a production-related, economic non-governmental organization. In this chapter, the term NGO is used loosely to cover non-profit research, advocacy, or support organizations at the international, national, and local levels.

It is almost easier to define NGOs in the negative. They cannot sign treaties banning the export of hazardous wastes. They cannot tell recipients of World Bank loans that they must consider whether the funds improve the status of women in their countries. They cannot pass legislation that sets targets to reduce carbon emissions. They cannot issue an order to a multinational corporation to redesign its processes in the next few years to eliminate the use of chlorofluorocarbons. But they can lobby for all these actions to be

Today, there is a chance for all of these groups to work together on a broad front. . . . The many issues of the many non-governmental organizations come together in the one issue of sustainable human progress. . . .

With the advice and watchdog role of the NGOs, the chances for real change will be greatly increased. We inserted a section in our final chapter entitled 'Making Informed Choices' in which we call for a strengthening in the roles of the scientific community and non-governmental organizations to help governments do just that: make better-informed choices of options.

I therefore challenge the NGO community tonight to take advantage of this atmosphere and to test our report and government reactions to it to the utmost. . . .

Are the governments and international organizations publicly committing themselves to the ideals of sustainable development and privately going on with business as usual? Judge them; prepare report cards. We have of necessity presented a general case for sustainable human progress, for planetary stewardship for the future. But this is meaningless unless sustainable development is woven into the fabric of national policies and laws. I challenge all national NGOs to work with governments in preparing national strategies for sustainable development and national audits of environmental resources and their uses.

Source: Excerpted from the James Marshall Memorial Lecture by Gro Harlem Brundtland, New York, 19 October 1987.

taken. And sometimes they provide the services governments are unable to.

Estimates of the numbers of NGOs are equally difficult to make. *Time* magazine recently reported that world membership just in environmental groups had risen from 13.3 million in 1988 to 15.9 million in 1989, though how they arrived at these numbers is anybody's guess. Even on a regional basis, researchers hesitate to count heads. Mary Helena Allegretti of the Institute of Amazonian Studies in Brazil, for example, told a 1989 conference that 'no detailed and accurate listing of environmental NGOs in Latin America is available and it would be risky to present disjointed and incomplete data'.

Whatever they are, and however many exist, they have clearly gained official recognition over the last twenty years as a force for change. Some 900 NGOs have 'consultative status' with the United Nations, for example, a list revised every two years. It is not considered unusual now to see someone from the International Institute for Environment and Development (IIED) or the

Environment Liaison Centre International (ELCI) at meetings of, for example, the UN Environment Programme, prowling the halls to lobby official delegates on the best way to combat desertificaton. And in 1976 the UN established a Non-governmental Liaison Service to promote development education among industrial-country NGOs.

The World Bank has long been a target for NGOs concerned about the lack of progress in alleviating poverty and, more recently, about the environmental impact of projects funded by development banks. In 1982 a World Bank–NGO Committee was created to provide a regular forum where these concerns could be aired. Although some NGOs dismiss this as just window-dressing, a number of people widely respected in the NGO community still sit on the Committee, and believe it is at least a useful point of entry for influencing Bank policy. The Committee helped prepare a 1988 directive to all Bank staff that said as a 'matter of Bank policy' they were to develop contacts and collaborate with NGOs.

In the environment field, the last twenty years has seen a considerable leap in the credibility, sophistication, and influence of NGOs. In June 1972 Stockholm was the site of not only the UN Conference on the Human Environment but also an Environment Forum, a People's Forum, and a camp full of devoted environmentalists at an airfield outside the city. Those protesting the main deliberations heard poems about whales and appeals from Hopi, Mohawk, Navaho, and Brazilian Indians. They toured the streets in an old bus made to look like a whale. And they were carefully kept away from the UN conference buildings by worried security officers. A few long-established NGOs scraped together a statement to the conference under prodding from Margaret Mead and Barbara Ward. Meanwhile, the Environment Forum became a jumble of protests about Vietnam and freedom fighters around the world. The *Guardian* in London reported the reaction of two British schoolchildren to this scene: 'The Forum looks like a school open evening without any parents.'

In 1990 environmental NGOs are quite likely to sit down with top government officials, especially in the industrial countries, on a regular basis. And the person across the table is now sometimes particularly familiar with NGO concerns: French Minister of the Environment Brice Lalonde was one of the founders of Les Amis de la Terre (Friends of the Earth in France). The Administrator of the US Environmental Protection Agency (EPA), William Reilly, was for many years head of the Conservation Foundation/World Wildlife Fund-US, an influential research and funding group on environment and nature conservation issues.

Although not in government, legislators with strong environmental backgrounds are also proving to be important allies for NGOs. In Brazil, for example, a few years ago Fabio Feldman was the attorney for the victims of pollution in Cubatao; in 1988, as a recently elected Federal Deputy, he helped draft into the new constitution a number of key paragraphs based directly on the report of the World Commission on Environment and Development (see Chapter 3).

With the rise of green parties, particularly in Europe, the access of NGOs to the political and policy-making process is likely to become even easier. This could be particularly striking in Eastern Europe, where environmental groups have been the lightning rods for public discontent about a variety of societal ills in the last few years. Yet the condition of the environment is truly one of their greatest concerns.

In February 1990, 100,000 people in Czechoslovakia, Hungary, and Austria formed a 100-mile human chain to protest against a proposed hydroelectric project. The preceding month, a delegation from the Soviet Union that met with the US EPA on areas of possible co-operation included Maria Cherkasova of the Social-Ecological Union, a consortium of 120 environmental groups in 260 towns. As these nations move closer to what the Brundtland Commission called the first prerequisite for the pursuit of sustainable development—a political system that secures effective citizen participation in decision-making—environmental NGOs can be important forces for change.

Communication and co-operation among environmental NGOs has also improved in the last decade, in part through the use of computer technology. Many offices now have sophisticated computer networks and facsimile links, so that news of important legislation, of visits by key environmentalists, or of ecological disasters that are not widely reported can move around the world in a matter of minutes.

The Environment Liaison Centre International in Nairobi has built up an extensive database on NGOs over the last decade, and can link international groups with grass-roots activists working on a wide range of issues. The *NGO Networker*, a newsletter from the World Resources Institute, also helps groups stay informed. Co-operation among groups was helped recently as plans were made to celebrate Earth Day 1990 around the world.

Groups working on the many issues surrounding more traditional Third World development concerns are even more varied than those in the environment field. Here, the major change in recent years is the explosion of local groups and the growing recognition of their value. Numbers, again, are nearly impossible to gauge. Alan Durning of Worldwatch Institute recently wrote about their membership running in 'the hundreds of millions'. He cited, for example, the Sarvodaya Shramadana village awakening movement in more than 8,000 Sri Lankan villages, some 12,000 independent development organizations in India, close to 25,000 women's groups in Kenya, and 1,300 neighbourhood associations in São Paulo, Brazil.

In a *Worldwatch Paper* on grass-roots groups, Mr Durning captures well their origin and impact:

> In the Third World, the birth of modern grassroots movements is a dramatic departure from historical precedents. . . . traditional tribal, village, and religious organizations, first disturbed by European colonialism, have been stretched and often dismantled by the great cultural upheavals of the twentieth century: rapid population growth, urbanization, the advent of modern technology, and the spread of western commercialism.
>
> In the resulting organizational vacuum, a new generation of community and grassroots groups has been steadily, albeit unevenly, developing since mid-century, and particularly over the past two decades. This evolution is driven by a shifting constellation of forces, including stagnant or deteriorating economic and environmental conditions for the poor, the failure of governments to respond to basic needs, the spread in some regions of new social ideologies and religious doctrines, and the political space opened in some countries as tight-fisted dictatorships give way to nascent democracies. In contrast to traditional organizations and mass political movements, this rising tide of community groups is generally pragmatic, focused on development, and concerned above all with self-help.

This chapter could not begin to cover the many thousands of NGOs that exist. Instead, it includes profiles of a few organizations. Some are famous around the world, like Greenpeace, and most are well known at least in the environmental community. Some groups focus on a single issue—tree-planting and pesticide use are the ones discussed here—or a particular audience, business students in this case. Others take on the full range of sustainable development issues at the national level. Still others serve as clearing-houses for NGOs in their regions or around the world.

Ten groups are included here. Just ten among hundreds of thousands. But these ten give some hint of the range of potential for individuals to get involved in shaping our common future.

* * *

One of the better-known Third World NGOs is the Green Belt Movement in Kenya. This community tree-planting campaign began in 1977 when a tireless campaigner for women's rights and for the environment, Wangari Maathai, joined the National Council of Women of Kenya (NCWK). She had earlier tried unsuccessfully to start a programme called Envirocare to plant trees in one Nairobi neighbourhood. With the NCWK's backing, however, the idea began to take hold as part of the *harambee*—'let's all pull together'—movement that was popular then.

At the first tree-planting ceremony, on World Environment Day in 1977, the dedication that was read contained two of the keys to Green Belt's success—an understanding of the pressures on the nation's land and an appreciation of the impact one person can have:

> Being aware that Kenya is being threatened by the expansion of desert-like conditions, that desertification comes as a result of misuse of land and by indiscriminate cutting-down of trees, bush clearing and consequent soil erosion by the elements; and that these actions result in drought, malnutrition, famine and death; we resolve to save our land by averting this same desertification by tree planting wherever possible.

> In pronouncing these words, we each make a personal commitment to our country to save it from actions and elements which would deprive present and future generations from reaping the bounty which is the birthright and property of all.

Soon the Nairobi Primary School agreed to establish a nursery and a belt of trees around its borders, and the campaign took off. Members were encouraged to find public areas where 1,000 seedlings could be planted to form tree belts. The Ministry of Environment and Natural Resources provided seedlings free of charge until the demand became so great the government could not afford to continue. At that point, Green Belt started paying a few cents for each seedling women could provide to local campaign-run nurseries, providing a valuable supplement to the subsistence farming so many rely on.

By 1989 the movement claimed to have established 670 community tree nurseries, planted some ten million trees in 1,000 belts of at least 1,000 trees each, and inspired people to plant another 20,000 'mini-green belts' on private property. Over the years, 50,000 Kenyan women have been involved in the campaign. Several films have documented the campaign's successes, and Professor Maathai, as she is known throughout Kenya, has been added to the UN Environment Programme's Global 500 Roll of Honour for Environmental Achievement. In 1989 she travelled to London and joined Mother Teresa in receiving 'Women of the World' awards from WomenAid.

Although booklets from the movement's early days talk about the objectives of promoting tree-planting through community participation and convincing planters to take care of the trees, within a few years a broader goal of public education about sustainable development became clear to the group: 'The Green Belt Movement is more than just planting trees. It is an educational experience aimed at changing mental attitudes towards the environment through greater awareness, understanding and appreciation.'

The main objective, a recent brochure notes, is to raise the consciousness of people to the level where they will do the right things for the environment 'because their hearts have become touched and their minds convinced'. Green Belt's goals include providing a source of income for women who produce seedlings and monitor the nurseries, and building confidence in local people who often feel overwhelmed by outside experts rushing in to tell them what is best for them.

The group aims to help women become creative, effective, assertive leaders and partners in development. Few women have been elected to Parliament in Kenya as yet, and both women and small farmers—often an overlapping group—are marginalized segments of the society. As the Brundtland Commission noted, 'social and cultural factors dominate all others in affecting fertility. The most important of these is the roles women play in the family, the economy, and the society at large.' The Green Belt Movement includes the need to address the population question as one of its objectives, and its efforts to improve the status of women in Kenyan society offer some hope of slowing one of the highest population-growth rates in the world.

In a 1988 booklet called *Sharing the Approach and the Experience*, Professor Maathai described five sources of Green Belt's success: the campaign recognizes the importance of immediate benefits, has had adequate finances, enjoyed freedom of operation, is run by dedicated volunteers and staff, and—last but not least—is located in a country that has a great demand for trees. Yet she notes that 'it is easier to describe the Movement's strategy, even to show how it works, than to explain *why* it works. People want to understand it as they would understand a foolproof recipe for a delicious dish. But there is no recipe.'

The ingredients for success in another group, the Pesticide Action Network (PAN), include good communication and autonomy. PAN was formed in 1982, when key activists campaigning against what they call the 'circle of poison' met in Malaysia at a conference organized by Sahabat Alam Malaysia (Friends of the Earth, Malaysia) and the International Organization of Consumers Unions. In just eight years, the network has grown to more than 300 organizations in some fifty countries. They stay in touch through six regional centres around the world.

No one individual speaks for the entire coalition, and members pursue whatever projects their own groups decide are appropriate. Some concentrate on lobbying; others tell farmers about alternative ways to control pests. In general, the members oppose the unnecessary use and the misuse of pesticides, lobby for immediate notification by governments of bans or restrictions on any pesticide, and urge greater reliance on safe pest-control methods, such as integrated pest management. They also stress that pesticide use is but one part of larger structural problems in developing countries' agricultural sectors; efforts must also be made, PAN maintains, to help farmers reduce their dependence on agricultural technologies pushed on them by international aid and financial institutions.

According to the coalition, pesticide sales have quadrupled since 1960, and more than four billion pounds of these chemicals are applied each year. The North American regional centre reports that the World Health Organization estimates more than a million people are poisoned annually by pesticides; 20,000 of them die as a result.

To combat these tragic occurrences, and to help farmers get off the 'pesticide treadmill', in 1985 PAN launched the Dirty Dozen Campaign. This put the spotlight on particularly hazardous chemicals, some of which were already banned in industrial countries but still exported to the Third World. The coalition claims to have made considerable progress: 'After four years of the Campaign, ten of the Dirty Dozen are the most widely banned or severely restricted pesticides in the world. Following successful efforts to phase out the production of 2,4,5-T and limit the distribution of chlordimeform, many Dirty Dozen campaigners are focussing on the extremely toxic herbicide paraquat.'

Several times a year the six regional co-ordinators sit down together while they are all at a UN meeting or some conference on agriculture. They also rely on conference telephone calls and computer links to keep up to date with developments in the various regions. Thus PAN provides a good example of how NGOs that focus on a single issue can make the best use of limited time and resources.

As mentioned earlier, NGOs sometimes focus on a single audience rather than a single topic. One obviously key group for our common future is youth—the ones who in a decade or two will have to deal with this generation's failures to move quickly enough. The International Association of Students in Economics and Management (AIESEC, from the group's name in French) is an interesting youth organization formed more than forty years ago. In 1988 this 50,000-member organization launched a two-year seminar series entitled 'Our Common Future: The Challenge of Cooperation', in direct response to the Commission's report and to the UN Secretary-General's call for youth education. The effort has received support from Shell International Petroleum of Great Britain, and sponsorships from the World Bank and IBM-UK.

The series of more than 100 seminars has focused on sustainable development in relation to business, in order to prepare 'today's students to be more responsible future managers'. AIESEC's headquarters is in Brussels, but it has chapters in sixty-nine countries, on the campuses of 650 universities, most of which have found the sustainable development theme of great interest and relevance.

Seminars have been held over the two years of the series on 'Changing the Mind and Heart of Business' in Ireland, for example, on 'Human Settlements—The Urban Challenge' in India, and on 'Latin American Perspective Towards the Year 2000' in Colombia.

With the help of the US Information Agency, business students in West Africa were able to talk with agricultural specialists in Washington in an interactive television seminar on food security in the developing world.

In August 1990 some 400 AIESEC members will meet in Tokyo for a World Theme Conference that pulls together the whole series. The meeting's title: 'A Youth Perspective on Sustainable Development—The Role of Economics and Education.' The aim: to have delegates set both short- and long-term goals for youth around the world. Delegates will be drawn not only from AIESEC chapters but also from other youth groups, NGOs, and government agencies.

As noted in Chapter 1, when *Our Common Future* was released in London in 1987, Mrs Brundtland presented copies of the book to twelve young people, saying: 'Securing our common future will require new energy and openness, fresh insights, and an ability to look beyond the narrow bounds of national frontiers and separate scientific disciplines. The young are better at such vision than we, who are too often constrained by the traditions of a former, more fragmented world.' With this global seminar series, AIESEC has made a bold attempt to tap that energy and openness, to capture those fresh insights.

*   *   *

During the last decade, numerous groups have broadened their focus beyond a single topic or audience. Several have embraced the full range of issues covered by the term 'sustainable development', trying to avoid the fragmented concerns that Mrs Brundtland referred to in her 1987 speech in New York.

One good example of this at the most local level is the Sustainable Development Communications Project (SDCP) in Vancouver, Canada. Its three founders—the Outdoor Recreation Council of British Columbia, the BC Forestry Association, and the BC Wildlife Federation—decided in March 1988 that the province needed a central resource-and-information centre on sustainable development. They were inspired by *Our Common Future*, they report, and by the wealth of activities to be reported on in British Columbia.

This province was the last in Canada to set up a Round Table on Environment and Economy (see Chapter 7), yet before it was even appointed in January 1990, local task-forces had formed in towns throughout British Columbia. The issue of logging is naturally a major concern in the area, and these groups bring together people from industry, tourism promotion, resource development, and environmental NGOs to consider how best to meet their needs in a sustainable way.

The SDCP office has been tracking these provincial developments in a useful four-page monthly news-letter called *Focus B.C.*, which also lets British Columbians know about relevant national publications and events. Following an initial circulation of 500, the news-letter was reaching 1,400 people within a year, a good sign of the need for such a clearing-house. The office has even received letters from Poland and Africa requesting follow-up information, so *Focus B.C.* is reaching much further afield than the founders expected. With a staff of just three and a tiny budget, the success of this group demonstrated the thirst for information at the local level. In February 1990 the new provincial Round Table agreed to fund production and distribution of the news-letter.

In the United States, the major group that works to pull together all the stake-holders in our common future is the Global Tomorrow Coalition (GTC). Founded in 1981, GTC now speaks for more than ten million Americans in the 115 organizations that belong to the coalition. The groups included run the gamut of those working on environment and development issues—from the Acid Rain Foundation to Zero Population Growth. With a small staff based in Washington, DC, the Coalition has spent the last several years explaining the concept of sustainable development in a country where the term was rarely heard.

A comprehensive 350-page *Citizen's Guide to Sustainable Development* provides an overview of priorities and progress. 'A first step toward action', notes GTC President Don Lesh, 'is a better understanding of the complex interrelationships between problems of environment, development, population, and resources.' The book is designed to help people see those links, and to give them information on how to get involved locally or nationally. The Coalition has also developed education materials on such topics as biological diversity and population that have been used by more than 20,000 teachers and children in elementary and high schools across the country.

One process that has been remarkably successful for GTC in broadening public understanding of sustainable development is Globescope, a series of public assemblies that have brought together the subject's fragmented concerns. GTC Vice-President Diane Lowrie calls Globescope 'the single most ambitious grassroots program on global issues in the country today'. At the first one, in Portland, Oregon, in 1985, some 800 participants spent five days discussing sustainable development. More to the point, they spent the first day learning about the issues this concept encompasses, and then three days in small working groups coming up with specific actions that they as individuals should—and would—take to address the problems.

Trained 'facilitators' helped participants sort through the range of opinions on actions needed and come up with a priority list at the end of each session. On the final day of Globescope, the group as a whole reviewed all the suggestions and drafted an Action Plan with sections on changes in public policy, education, communication, and personal actions. The plan was accepted unanimously. Jeff Strang, the publications co-ordinator in Portland, found that the 'process was an education in consensus and direct democracy, comparable in spirit to the difficulties and triumphs experienced by our nation's Constitution drafters nearly two hundred years ago'.

Thousand of Americans have participated in Globescopes since 1985—from Oshkosh, Wisconsin, to Sun Valley, Idaho. The largest one was held in Los Angeles in November 1989 and co-sponsored by the Centre for Our Common Future. This brought together some 1,500 people to help 'set a new US agenda for environment and development in the 1990s'. The initial two days of the meeting consisted of public hearings on the report of the World Commission on Environment and Development. Testimony from forty individuals was heard by a specially convened Council on Sustainable Development that included former President Jimmy Carter, former members of the Commission, leading industrialists, and the heads of environmental, youth, and religious organizations.

The meeting then broke up into working groups that drafted, as in other Globescopes, detailed recommendations for citizens' groups, business and trade unions, teachers and young people, science and technology, governance, and individual responsibility. They run from the cosmic (improve quality of life and provide universal health-care by the year 2000) to the pedestrian (a pledge to bicycle or walk on trips of less than two-and-a-half miles by the year 1990). GTC used the results of this process as the basis for *Sustainable*

Development: A US Response, which was submitted to the President and Congress for Earth Day 1990.

In a message recorded for the opening session of the meeting in Los Angeles, Mrs Brundtland noted that 'I cannot overemphasize how much I welcome the leadership of the United States in this international effort, and—for the same reasons—how strongly I feel that the Globescope Pacific Assembly is an event not only of national, but international importance'. The *US Response* developed by the Global Tomorrow Coalition puts a powerful tool in the hands of all Americans who want to press their government to exercise that leadership role.

In Australia, the coalition approach of GTC was mirrored in 1989 in 'One World or . . . None, A Campaign for Global Change'. Based in the offices of the Australian Council for Overseas Aid, the campaign reached out to the many constituencies already working on sustainable-development issues. Its aim was to focus 'on the links between environmental destruction, crippling Third World debts, abuse of human rights and a mushrooming arms trade'. Although this was supposed to be a six-month effort, through November 1989, the response was so positive that the organizers decided to continue a little longer, at least through the federal election in spring 1990.

Plans for this campaign were well under way when the organizers heard about the global television-and-radio broadcast, 'Our Common Future', on 3 June 1989 (see Chapter 6). Working with the broadcast team in Washington, they arranged the launch of the whole campaign for the same weekend. A toll-free number was set up in Australia, as it was in many countries carrying the broadcast, and the response was 'overwhelming and heartwarming', the organizers report. Their volunteer phone-tenders took 10,000 calls over forty-eight hours, from people wanting to know what they could do in their own homes, their offices, their communities to help. In a country of just over sixteen million, 10,000 calls is quite a response.

What callers received in response was a Campaigner Kit and a booklet called *Making a Difference* that was prepared for 'One World or . . . None'. Over the next few months, a steady stream of news-letters contained suggestions on green consumerism and investing, recycling, how to get people to sign the One World Declaration, and who to write to locally and nationally about specific issues. In addition, the campaign worked with the Commission for the Future on the *Personal Action Guide* published in late summer, an Australian adaptation of a popular US pocket-guide to living lightly on the earth.

The November news-letter listed the following as signs of the campaign's success: an indication from the Minister of Resources that a formal government response to *Our Common Future* should be prepared, a federal-government decision not to use imported rain-forest timbers in its construction projects, an Industries Assistance Commission study on the impact of government incentives on recycling, and growing consumer awareness about wasteful consumption matched by a growing commitment to recycle.

Each news-letter had a balance between environment and development concerns that reflected the Brundtland Commission's view of the importance of both these components of our common future. Indeed, the news-letters often pointed out that 'the environmental crisis cannot be solved without a solution to the poverty and debt crisis'.

Asked to comment for this book about recent developments in Australia, co-ordinator Janet Hunt wrote that the campaign 'contributed to a broadening out of concerns among the environmental movement itself. . . . There is certainly developing a much closer relationship in Australia between environment and aid and development NGOs. The media and some politicians are now also acknowledging that link.' The campaign's invitation to meet with Prime Minister Hawke in December made its success clear.

*   *   *

Helping groups to see the connections among their fragmented concerns is the goal of two important regional networks, each about a decade old. The African NGOs Environment Network (ANEN), based in Kenya, was established in 1982 by twenty-one Africans from local groups who met in Nairobi on the tenth anniversary of the Stockholm Conference. They decided that NGOs in Africa needed a strong voice to help them join governments and aid agencies as partners in development. Independent NGOs, Executive Director Simon Muchiru has noted, can 'fill the gaps in development projects with the aid and support of Governments, and in so doing much could be achieved to stem the degradation of the African environment'.

One of ANEN's first activities was a training workshop on the use and misuse of agricultural chemicals, which linked the group with PAN. In 1985 a bimonthly, bilingual magazine was launched— *Ecoafrica*. Studies have been done on grass-roots movements and on the state of the environment in Africa.

Since 1986 the organization has worked on a project funded by the UN Environment Programme to fight desertification. Through this, ANEN has been able to provide small grants to community groups in Botswana, Burkina Faso, Kenya, and Senegal. Plans have also been made to organize a network of African environmental journalists to raise awareness of the implementation of sustainable development in the African context.

A recurring concern for ANEN is the need to strengthen local NGOs. Many groups lack trained, experienced staff, not to mention equipment for them to work with. In fact, ANEN's *Programme of Action 1988–1990* includes an appeal for a computer and a photocopying machine—essential supplies if they are to establish an environmental library. The staff remains quite small, constraining ANEN's ability to tackle new issues. Funds that can help train people and let groups be more effective are constantly needed.

Despite all these obstacles, ANEN Chairman Jimoh Omo Fadaka wrote in January 1989 that he was happy to report the group had received institutional support and legitimacy: 'This recognition is important because for a long time African governments have viewed NGOs as anti-government organisations. It appears that there has been a change of attitude, which has been brought about by the changing focus of development activity in the continent. NGOs have become formidable partners in progress and embodiments of revitalising ideas and practices.'

In the Philippines, formidable partners have also joined together. A small office in Manila with the unwieldy title of the Asian NGO Coalition for Agrarian Reform and Rural Development, better known as ANGOC, claims to be one of the pioneers of networking in Asia. The coalition was established in 1979 when representatives of several groups met at a workshop in Bangkok to make recommendations for a forthcoming UN conference on agrarian reform. They found they had a great deal of information to exchange and experiences to share.

A decade later, twenty groups belong to ANGOC as national NGOs or research institutes, from Bangladesh, India, Indonesia, Pakistan, the Philippines, Sri Lanka, and Thailand. Included, for example, are the Association of Development Agencies in Bangladesh and the Consortium on Rural Technology in India. A number of these groups are themselves coalitions, so more people are represented than a figure of twenty members would imply. Many of them work to increase the access of the poorest of the poor to credit, agricultural supplies, and other basic services.

For the first few years of its existence, ANGOC focused on building a basic network through workshops and public education campaigns. Executive Secretary Edgardo Valenzuela notes that 'NGOs in Asia have come to the realization that the task of building a more just and a more human society is not a monopoly of government but rather a collective task for all concerned citizens'. To ease that task, the coalition promoted information exchanges and meetings with governments and UN offices to acquaint them with members' strengths, and with their needs.

As the group became better established, studies were undertaken on rural poverty alleviation, marginalization of farm workers, food issues, and the role of NGOs in rural development. Case studies have been published on successful programmes in most of the region, and a quarterly journal looks into such issues as social forestry, co-operation among groups, and the need to include rural women in development programmes. Since World Food Day was launched by the UN Food and Agriculture Organization, ANGOC has used this annual event as a focus for local activities.

In mid-1989, as part of a major effort to reorient its programmes towards sustainable development, ANGOC co-hosted a meeting in Manila that combined their experiences in development with their growing concerns about the state of the environment. Thirty-one NGO leaders from around the world attended this strategy session. The resulting Manila Declaration on the importance of authentic development has received wide circulation since it was issued in June (see Box 4–2).

In addition to defining people-centred development, the Declaration noted that a transition period lay ahead while NGOs worked with allies in governments and donor agencies to transform the existing system. During this transition, efforts must be made both to

Box 4–2.   The Manila Declaration on People's Participation and Sustainable Development

Current development practice is based on a model that demeans the human spirit, divests people of their sense of community and control over their own lives, exacerbates social and economic inequity, and contributes to destruction of the ecosystem on which all life depends. Our work with grassroots communities brings us into daily contact with the results of this development.

Furthermore, we are concerned that foreign assistance, particularly debt financing, too often contribute more to the problem than to its solution. It places the initiative and responsibility in the hands of foreigners rather than in the hands of the people. . . . It results in the imposition of policies intended to facilitate debt repayment, orienting the national economy and its resources to the needs of foreign consumers, at the expense of the poor and the environment.

There is current need for a fundamentally different development model based on an alternative development. Authentic development enhances the sustainability of the community. It must be understood as a process of economic, political and social change that need not necessarily involve growth. Sustainable human communities can be achieved only through a people-centered development.

A people-centered development seeks to return control over resources to the people and their communities, to be used in meeting their own needs. This creates incentives for the responsible stewardship of resources that is essential to sustainability.

A people-centered development seeks to broaden political participation, building from a base of strong people's organizations and participatory local government. It seeks the opportunity for people to obtain a secure livelihood based on the intensive, yet sustainable, use of renewable resources. It builds from the values and culture of the people. Political and economic democracy are its cornerstone.

stop damaging the environment and people's prospects for socio-economic development, and to create alternatives. Eight essential steps during the transformation were identified: redefining participation, opening access to information, building inclusive alliances, reducing debt dependence, reducing resource exports, strengthening people's capacity for participation, creating demonstrations of self-reliant communities, and creating national and international monitoring systems.

At ANGOC, one theme that emerges in their plans for their second decade is the importance of strengthening the capacities of their members. As the African network's appeal for basic supplies

It seeks to build within people a sense of their own humanity and their links to earth, its resources, and the natural processes through which it sustains all life. The relationship of the people to the land is of particular importance. Alienation from the land creates a symbolic alienation from community and from nature. . . .

Three principles are basic to a people-centered development.

1. Sovereignty resides with the people, the real social actors of positive change. . . .

2. To exercise their sovereignty and assume responsibility for the development of themselves and their communities, the people must control their own resources, have access to relevant information, and have the means to hold the officials of government accountable. . . .

3. Those who would assist the people with their development must recognize that it is they who are participating in support of the people's agenda, not the reverse. The value of the outsider's contribution will be measured in terms of the enhanced capacity of the people to determine their own future. . . .

Many of the changes called for by a people-centered development present a fundamental challenge to well established interests. A call for ideal changes would be unrealistic, were it not for the depth of the crisis of deepening poverty and environmental destruction that now confronts human society. The future of all people depends on a basic transformation in thought and action, leading people to re-discover their essential humanity and to re-create their relationships with one another and among themselves and their environment. It is pragmatism more than idealism that makes the change possible.

Source: Excerpted from Declaration adopted at the Inter-Regional Consultation on People's Participation in Environmentally Sustainable Development, Makati, Metro Manila, Philippines, 6–10 June 1989.

makes clear, NGOs in the Third World must often struggle to get basic information out, and have little time and money to do research. Their offices are a far cry from the plush surroundings of groups in Washington or London. Which makes it all the more important that they receive support from their wealthier industrial-country counterparts. And that they join together in coalitions like ANGOC, where they can share their experiences, their expertise, and their resources.

*　　*　　*

The ANGOC meeting on environmentally sustainable development was organized in co-operation with the Environment Liaison Centre International, the leading coalition of NGOs that are working towards sustainable development. ELC was established in 1974 following discussions at the Stockholm Conference; in 1987 'International' was added to its name to reflect the global membership. And paralleling the growing understanding of the links needed among groups, ELCI now also identifies itself as the Global Coalition for Environment and Development.

ELCI, based in Nairobi near the UN Environment Programme headquarters, has 340 members but is in contact with more than 8,000 NGOs around the world. More than half the member organizations are located in developing countries, and two thirds of the Board of Directors that is elected annually are from the Third World. The office works to increase communication between private groups working on similar issues, and helped form, for example, the Pesticide Action Network. In fact, ELCI serves as the PAN regional centre for English-speaking Africa.

'The human community and the earth on which it draws its sustenance' are at the centre of ELCI's concerns, according to its strategy for 1989-91. This leads it to two overall goals: 'to ensure enduring livelihoods for the majority of the people, mainly in the present economic South, and also indigenous and other deprived people everywhere in the short and medium term; and to seek sustainable approaches to development for the global community based on equitable access to the resources and respect for nature.'

ECOFORUM, a bimonthly magazine in English, French, and Spanish (and soon in Arabic), lets members exchange information and debate key issues. Five programme areas are the main focus of ELCI's work—women, environment, and development; food security and forestry; energy; international economic relations; and industrialization and human settlements. In 1989 the centre launched WEDNET, the Women, Environment and Development Network, with funding from the International Development Research Centre of Canada. This new computerized link between researchers will look at women and the management of natural resources in Africa, and could replicate PAN's success in raising visibility of a long-neglected issue.

While members around the world develop their own position papers and programmes on all these vital areas, collaboration through ELCI can sometimes magnify their voices. A statement on climate change, for example, was drafted by some NGO representatives who got together while they were attending official conferences on that topic in Toronto and Hamburg in 1988. After being circulated widely throughout the ELCI membership, a final NGO statement was submitted to the UN Environment Programme Governing Council in May 1989, with endorsements from eighty-nine organizations in thirty-two countries.

To aid Third World grass-roots groups, ELCI has a Small Grants and Local Initiatives Support Fund, which can provide up to $5,000 to NGOs; the fund is financed for the most part by the government of the Netherlands. In 1988 the recipients of these small grants included a meeting of people displaced by large dams in Brazil and environmental-education campaigns in India and Sri Lanka.

The report of the Brundtland Commission has been the topic of much discussion among NGOs around the world since the day it was published. To consider its implications for this community, ELCI arranged a December 1988 meeting in Tabarka, Tunisia, called ' "Our Common Future"—Making It Happen'. The title neatly summarized a question often heard about the Commission's report: where do we go from here?

ELCI's Executive Director, Shimwaayi Muntemba, opened the meeting with her summary of the report's impact: 'When the dust settles, the great contribution of *Our Common Future* will be seen to lie in pushing the environment/development linkages into political and governmental arenas and thus bringing legitimacy, and immediacy, to the beliefs and work of the ELCI constituency.'

Not everyone agrees with that assessment. The two ends of the spectrum of NGO opinion on the Brundtland Commission met in Tabarka, as conference highlights prepared by Lloyd Timberlake of IIED indicate:

> The major division which emerged on the first day was between those who saw *Our Common Future* as the most important opportunity for environmentalists since the 1972 Stockholm Conference on the Human Environment, and those who saw it as 'a threat' in that it legitimises rapid economic growth under the phoney rubric of 'sustainable development.' But these extremes are not fair, in that many speakers saw it as both a document in part 'profoundly wrong,' and at the same time an effective lever which relatively weak NGOs can use to move governments.
>
> *Our Common Future* will force a shake-out in the NGO movement. Those opposed will have to explain, practically and specifically, how one can resist economic development and growth while at the same time being in favour of the elimination of severe poverty and the

meeting of basic needs in a world of rapid population growth. Those in favour will have to explain, practically and specifically, exactly what sort of economic development they are espousing, how it can be had in the North in ways which also contribute to the 'development' of the South.

Getting down to practical specifics is the challenge facing ELCI and its members in the nineties. As part of its current three-year programme, and as an outcome of the discussions in Tabarka, ELCI is focusing on strengthening grass-roots activities—best done through a new regional focus, it believes. The organization thus has the difficult task of both representing all its members at international meetings, and working to decentralize its response to the issues. Regional consultations, like the one organized in Manila with ANGOC, will no doubt help their efforts to accomplish this.

*   *   *

As indicated earlier, it is impossible to cover here even a representative sample of the many kinds of non-governmental organizations in existence today. But it would also be impossible to talk about groups that are a force for change without including Greenpeace—the fastest-growing environmental group in the world. It started with about twenty people in 1971; it claims to have more than four million supporters in 1989. Greenpeace now has thirty-three offices in twenty-three countries, plus a World Park Base in Antarctica.

David McTaggart, one of those first twenty supporters and now chair of Greenpeace International, was not even at the Stockholm Conference, a magnet for most environmentalists in 1972. Four days before that conference opened, as official delegates, assorted Friends of the Earth, and thousands of devoted environmentalists headed to Sweden, Mr McTaggart and a two-person crew in the *Vega* crossed an invisible line in the ocean near Moruroa Island, in the South Pacific, which the French government had declared off limits. The French had been using this area for nuclear tests since 1966.

The *Vega* stayed around Moruroa long enough to delay the testing, get rammed by a French frigate, and arouse public opinion about atmospheric nuclear testing. When the boat returned a year later to protest against the next tests, David McTaggart and Nigel Ingram were severely beaten up by sailors from a French military boat—an act captured on film by Anne-Marie Horne, who was able to hide the camera during the chaos of the fight. By November 1973, after howls of protest when the photos were released, the French government announced that its nuclear tests in the future would all be conducted underground. Thus was born the Greenpeace strategy: put your life on the line in defence of the earth, and make sure there is film in the camera.

Greenpeace's base in the Quaker tradition of 'bearing witness', of letting your life speak for your values, may be less well known than its daredevil stunts. The group was started in Vancouver in 1971 by a few Canadians and expatriate Americans who were searching for a non-violent way to protest against US nuclear testing on the Alaskan island of Amchitka. One of the group, Irving Stowe, was a Quaker; another couple, Jim and Marie Bohlen, had grown up in a part of Pennsylvania where many Quakers live. To these three, sending a boatload of people into an area where a nuclear bomb was to be exploded was a perfectly logical way to demonstrate, by their presence as witnesses, their sense of the truth.

Greenpeace volunteers have been on many front lines and front pages since then. Standing on ice floes in north-eastern Canada blocking the way of seal hunters. Plugging up underwater pipes that are discharging toxic wastes into rivers and harbours. Handcuffing themselves to drums of toxic wastes that are to be dumped in the sea. Dangling off cliffs and bridges around the world with signs about whaling, about destruction of the ozone layer, about toxic chemicals, about nuclear power, about Antarctica, about whatever else Greenpeace thinks the world should hear its view of the truth on.

These attention-grabbers are called 'direct actions' in Greenpeace, and they take personal courage and strong convictions. The organization has also been known to engage in 'soft actions', such as turning up at the Environmental Protection Agency in Boston in a gorilla suit to hand out bananas with the message 'you must be bananas to be for incinerators'. (Greenpeace believes that government should stress waste reduction before states are encouraged to build plants to burn hazardous wastes.)

Behind these acts of bearing witness—and having a little fun sometimes—stands an organization that takes no government funds or corporate grants, but that had a world-wide budget estimated at $100 million in 1989. All the money comes from the four million, and counting, supporters. The French government could be called an indirect supporter, though for a tragic reason. Their second run-in with Greenpeace, in 1985, took the life of photographer Fernando Pereira. He was the only crew member on the *Rainbow Warrior* when the French sank the boat in the harbour of Auckland, New Zealand, before it was to lead a protest against underground nuclear testing on the Moruroa Atoll. Greenpeace won $8 million in damages from the French government for this act of sabotage, as well as literally millions of supporters.

The campaigns they support at the moment include the push to establish Antarctica as a world park; pressure to stop whaling by the last few nations that still allow it under the heading of 'scientific purposes'; an effort to ban shipments of hazardous wastes, which they claim force governments to choose between poison and poverty; and the dangers of having one quarter to one third of all nuclear bombs in the world on board ships.

The work of putting together an organization this large and far flung, and with strong-minded individuals who often have differing views of the truth, has at least bred a certain respect at Greenpeace for the difficulty of forging global responses to global problems. National sovereignty can be as much of an institutional obstacle for NGOs as for governments. David McTaggart now has to spend a quarter of his time dealing with this, he told the *New York Times*: 'The eternal struggle within the organization is to overcome the deep, inbred provincialism and nationalism and keep Greenpeace focused internationally. It's a big goddamned struggle, man, to keep everybody together.'

Greenpeace's strength so far has been in the industrial countries, though they also have offices in Argentina and Costa Rica. And they are just opening one in Brazil, which should ensure their visibility at the 1992 Conference on Environment and Development. As they now have consultative status with the United Nations, they will no doubt be seen both in the main conference halls, lobbying delegates on various campaign issues, and in the streets—with an inflated replica of a whale, perhaps, or a gorilla handing out bananas.

If Mr McTaggart had been at the Stockholm Conference instead of in the South Pacific, he would have heard Indira Gandhi speak in moving terms about the greatest polluter of all: poverty. Given the new office in Brazil and their intention of opening offices in other developing countries, it will be interesting to see what 'direct action' Greenpeace comes up with for a poverty campaign. People in the Third World have been bearing witness, and losing their lives, to that form of pollution for many years.

*   *   *

As official delegates and NGO representatives gather in Brazil in 1992, they might want to remember that David McTaggart was not at the 1972 conference. Nor, for that matter, was Brice Lalonde. In fact, the current French Minister of Environment did not help start Les Amis de la Terre until two years after the meeting in Stockholm. Delegates and dissenters in Brazil may reach a consensus on some plan to move the world closer to sustainable development. Governments could certainly take some important steps in 1992 to protect the atmosphere and the world's biodiversity. But in the world beyond the conference, more and more people are tired of waiting for things to get better. And for governments to help them.

In 1992 the McTaggarts and Lalondes of tomorrow may be in a boat on their way to cut the lines of any drift nets still stretched across the Pacific. Or organizing basic literacy and family planning courses for women in Kenya. Or starting a recycling campaign in West Germany. Or getting villagers in the hills of India to hug trees so that loggers will not chop them down. Or lining 50,000 people up along the Baltic Sea to bear witness to its polluted state. Or starting a door-to-door survey to figure out why an abnormal number of children in upstate New York have cancer. Or working with the pavement dwellers of Bombay to show their contributions to the city's economy. Or . . .

5

A Commitment by Producers and Consumers

The staff of an environmental group in Europe were taken aback recently when a visiting American unpacked a jogging bag and an aerosol can of deodorant fell out. They were surprised that the can did not have a prominent label saying it was 'ozone-friendly', meaning that no CFCs—chlorofluorocarbons—were used in it as the propellant. The environmentalists very carefully buy only aerosols that brag in large and bold type they contain no CFCs. What was their friend's excuse for not doing the same? The answer is simple, and something the jogger had no say in: nearly all aerosol cans manufactured in the United States have been 'ozone-friendly' since 1978.

This little story illustrates two points. First, even conscientious consumers have a hard time knowing everything that might influence their choice of product. And second, individual countries and even companies can take unilateral steps that affect the way the world does business.

The use of CFCs in aerosol cans was phased out (except for those needed in medical settings) in the United States over a decade ago after the first scientific reports about possible damage to the stratospheric ozone layer. Although US manufacturers protested vigorously at the time, they quickly came up with alternative propellants. But other governments did not introduce similar legislation, and aerosols in the rest of world stayed on the environmentalists' hit-list in the eighties.

During this time, international negotiations to phase out all CFC uses made slow progress until new scientific evidence underscored the danger (see Chapter 2). Indeed, *Our Common Future* contained only two paragraphs in 1987 on the damage to the earth's protective ozone layer that CFCs might be causing. The Brundtland Commission called on governments to ratify the existing convention and to develop protocols for the limitation of CFC emissions. Three years later, all that and more has happened. And European advertising campaigns are now built around ozone-friendly aerosols—a concept the public would not even have understood a few years ago.

In the United Kingdom, the quick turnaround no doubt received a boost when Prince Charles announced in February 1988 that he was banning CFC-propelled aerosols from his home. But it was solidly based on manufacturers' acceptance of strong scientific evidence about the escalating problem of ozone-layer depletion. Three days before Friends of the Earth were going to launch a consumer boycott of the offending aerosols in spring 1988, the eight leading UK manufacturers announced they would stop using CFC propellants by the end of 1989. The British Aerosol Manufacturers Association soon followed suit.

Now that it looks like the Montreal Protocol may be strengthened, to eliminate CFCs even earlier, manufacturers everywhere are racing to find substitutes and to try to earn the label of good corporate citizen. Du Pont, the largest CFC manufacturer in the world, announced in March 1988 that it would no longer produce these chemicals by the year 2000, phasing them out over twelve years. Their most recent investment in an alternative to CFCs is a $25-million plant in Texas to produce HCFC-134a.

Major users of CFCs in industrial nations have followed producers on to this bandwagon. General Motors announced that all its car dealers would have to recycle CFCs drained from air conditioners during servicing by 1991. The US communications giant AT&T, which uses three million pounds of CFCs annually as solvents during the manufacture of electronic equipment, pledged to halve its use by that same year, and to stop entirely by 1994. Nine of the biggest users even formed an Industry Cooperative for Ozone Layer Protection, to speed work on finding alternatives to using CFCs as cleaning solvents in electronics manufacturing.

The second atmospheric issue of great concern—climate change—seems so global that unilateral steps would accomplish little. Nevertheless, one US company, Applied Energy Services (AES) in Virginia, is trying to make a small contribution. In October 1988 this independent electricity producer announced it was joining the private development agency CARE in a $16.3-million project to plant fifty-two million trees in Guatemala. Since trees absorb carbon dioxide as they grow, the scheme would offset fifteen million tons of carbon dioxide that a new coal-fired AES power-plant is expected to generate over its lifetime. AES is contributing $2 million to the planting programme.

Guatemala was chosen because the country's deforestation problem is acute, and because CARE already had a project planned there but lacked funding. With the assistance of the Peace Corps and the Guatemalan forestry service, 40,000 farmers have been enlisted to plant trees. The AES initiative has been dismissed by some as being barely a drop in the bucket of changes needed to balance the world's carbon budget. Although it is a small step, its role as a symbol of a responsible approach by industry is important.

On another issue addressed by the World Commission—health— the steps taken by one corporation have been much more than symbolic. As *Our Common Future* pointed out, 'good health is the foundation of human welfare and productivity'. The Commission called for more research on the environmentally related tropical diseases that are the major health problem in the Third World. Merck & Company, a US pharmaceutical manufacturer, has acted on the results of such research.

In October 1987 Merck announced it would supply free, for all human treatments, the drug ivermectin, recently found to protect people from onchocerciasis (also known as river blindness). This goodwill gesture admittedly did not mean the company was giving up a lucrative market, as few of those who could use ivermectin would ever have bought it. The World Health Organization (WHO) estimates that eighteen million people in Africa and in parts of Latin America and the Middle East carry the parasitic worm that causes river blindness. More than 300,000 have already lost their sight. Ivermectin, developed in the battle against livestock parasites, has been found to halt the progression of onchocerciasis in infected individuals. Merck's decision to make it available at no cost will save the sight of thousands, and could help WHO eventually duplicate its triumph over smallpox: complete elimination of a disease from the earth.

A second health issue worrying a growing number of consumers is pesticide residues in food. Concern about the possibility that one class of chemicals (EBDC) causes cancer, and about the imminent move by the US Environmental Protection Agency to tighten controls on EBDC use, led the four major manufacturers to voluntarily suspend the use of EBDC on some fifty food crops in September 1989. Their initiative was hailed as responsible and unprecedented by one US environmental group, with the reservation that continued use of the chemicals on other food crops still posed a danger.

In Canada, the Crop Protection Institute released a booklet in 1988 that seems likely to translate into smaller profits for its members, a group of agricultural-chemical companies. The booklet encourages farmers to reduce excessive pesticide use, and indicates that the Institute hopes to halve the amount of pesticides washing off farm fields in the next decade.

Five small supermarket chains in the United States and Canada have taken responses to public concern about pesticides one step further: they announced in late 1989 a pesticide-reduction plan that will result in a complete halt to their sales of fresh fruit and vegetables treated with cancer-causing pesticides by 1995. The companies are small—Provigo Distribution Inc., ABCO Supermarkets, Raley's Supermarkets, Bread & Circus, and Petrini Supermarkets—and account for less than 2 per cent of the US and Canadian retail food industry. But as with AEG's tree-planting project, the example of corporate responsibility is important none the less.

*   *   *

Beyond these responses to specific environmental problems, some companies are starting to rethink the whole way they do business. It would be too much to say that sustainable development has been fully understood and warmly embraced by the major industries of the world. But some have taken steps in the right direction.

An encouraging number are following up on one or more of the six elements identified by the Brundtland Commission as a strategy for sustainable industrial development: to establish environmental goals, regulations, incentives, and standards; to make more effective use of economic instruments; to broaden environmental assessment; to encourage action by industry; to increase capacity to deal with industrial hazards; and to strengthen international efforts to help developing countries.

Some of these initiatives may reflect changes in who is running the shop. In an interview for this book about developments since 1987, former Commissioner William D. Ruckelshaus cited the new generation of business leaders as one sign of hope. These individuals realize that sound management means more than just meeting the letter of the law.

A recent speech by Thomas d'Aquino, President of the Business Council on National Issues (BCNI) in Canada, makes it clear which generation he identifies with: 'Twenty years ago, concern about the environment was mainly a preoccupation of the young. Our hair was longer then, we talked of new streams of consciousness, we questioned the impact of "mindless" technology.'

Indeed, in speech after speech over the last year, leaders of corporations—David Buzzelli of Dow Chemical Canada, E. S. Woolard of Du Pont, Roy Aitken of Inco, A. B. Cleaver of IBM, Richard Mahoney of Monsanto—have sounded more like Friends of the Earth than captains of industry. They claim to realize that enlightened self-interest is still good for business. And that the choice is not economic growth or ecological balance; it is both . . . or neither, as IBM puts it.

In response to *Our Common Future*, the UK office of IBM published a brochure called 'Sustainable Development' at a time when the term was hardly recognized in the US business community. In November 1988 IBM Europe donated equipment and software worth £3.6 million ($6.1 million) to the UN Environment Programme (UNEP), one of the largest corporate donations ever made to an environmental organization, IBM claims. And in Canada, the company has a National Environmental Co-ordinator.

In 1989 IBM Europe announced that a further $16 million was being donated to launch an IBM Scientific Centre in Bergen, Norway, in collaboration with academic and other industrial partners. The centre will be creating a database from projects in all 132 countries where IBM does business, using the six categories considered in the Brundtland report as common challenges—population and human resources, food security, species and ecosystems, energy, industry, and the urban challenge.

Why is this multinational corporation so keen to define itself as 'the company that cares about the future'? In the next decade, Nigel Corbally Stourton pointed out in an IBM brochure, 'we shall see more clearly the urgent need to manage the earth's resources on a global basis. Information technology will be of central importance in that scenario as providing the means of environmental modelling and resource management.'

Another multinational that believes it has a leading role to play is AB Volvo of Sweden. A special advertising supplement to *Time* magazine in November 1989 touted Volvo's long-standing claim to be the first manufacturer to declare a pro-environment stance. At the Stockholm Conference in 1972, the company said it did not 'wish to protect the auto at any price and under all conditions'. As chairman of the company since that proclamation, Pehr Gyllenhammar has gained an increasingly high profile on environmental issues.

Volvo, the largest company in Sweden, is working to eliminate its use of CFCs in car air-conditioning and foam seating at the earliest possible date, and lowering its solvent emissions during the painting of cars. The company has also created a Volvo Environment Prize. The first winner, in 1990, will receive about 1.5 million Swedish kronor ($245,000) for coming up with an innovation or discovery of significance to the regional or global environment.

Companies' attempts to develop broad-based strategies also include a wider use of environmental audits. The International Chamber of Commerce (ICC) defines these as 'a management tool comprising a systematic, documented, periodic and objective evaluation of how well environmental organisation, management and equipment are performing with the aim of helping to safeguard the environment by: (i) facilitating management control of environmental practices; (ii) assessing compliance with company policies, which would include meeting regulatory requirements'. In other words, a company checks on operating procedures to make sure corporate and legal requirements are being met and all systems are performing well.

These audits can be distinguished from environmental-impact assessments, during which companies, and perhaps government agencies, consider the likely effects on the environment of an operation while it is still in the planning stage. In Canada, for example, Inco Ltd.—the largest nickel-mining company in the western world—now routinely does an environmental assessment as part of the company's investment decision-making process. Assessments, or appraisals, have been standard in industrial countries for some time.

But audits are now also becoming important management tools. Some see them as key components of their attempts to reassure the public as well. Alex Krauer, chairman of Ciba-Geigy, pointed out in late 1987: 'It will no longer be sufficient to have good products, an efficient organization and a strong balance sheet. To be successful in the 1990s, you will have to win the acceptance of the people who live near your plants. People will have to believe in your professional competence and your ability to manage technology.' An audit can help that process. So can caring about the environment at highest corporate levels, as even Exxon has belatedly discovered: in January 1990 the company appointed Edwin Hess, a member of Exxon's staff for thirty-two years, as vice-president for environment and safety.

Former Commissioner Ruckelshaus is now Chief Executive Officer (CEO) of Browning-Ferris Industries (BFI), Inc., of Texas, one of the largest publicly held waste-management companies in the world. When he became CEO, he expanded the company's audits beyond the traditional hazardous- and medical-waste subsidiaries into all aspects of BFI's operations. He also pulled responsibility for these programmes into the central office, under the Vice-President for Environmental Affairs.

An auditor should be seen as just like someone who does financial audits, Mr Ruckelshaus maintains: a team player, not a 'gotcha' person. The goal is the efficient running of the corporation and the protection of public and environmental health, not just compliance with mandated standards. BFI has even financed consultants for community groups concerned about certain of the company's operations but unable to afford auditors. Having independent audits available as part of the public record may become more common as community groups and industry forge alliances, albeit uneasy ones.

Taking environmental audits one step further, British Petroleum (BP) now does 'issue audits'. These help management to see how the company is dealing with some environmental issues not normally thought of as BP's area, such as the destruction of rain forests. In 1988 BP became the first British company to receive the Gold Medal for International Corporate Environmental Achievement, an honour given annually by the New York-based World Environment Center.

Expanding the whole concept of environmental audits is an

important next step that a few firms have taken. Companies need to consider not just what happens to raw materials as they are turned into products, but the environmental impact of obtaining those materials. And not just how the company disposes of its wastes, but how consumers will be able to dispose of a product at the end of its useful life. Dow Canada, for one, refuses to sell its chemicals to customers it believes are not handling them properly.

In Europe, others have also started to take this broader view. In November 1989 the Round Table of European Industrialists decided to establish a permanent Working Group on Environment, with Volvo President Gyllenhammar as the chair. One of the group's first tasks will be to look at waste-disposal practices, particularly of chemical companies, and analyse changing corporate philosophies to see if more firms are looking at the 'cradle-to-grave' impact of their products.

A number of leading industrialists in Japan who are concerned about global issues have joined together as well. In December 1988 they set up the Global Industrial and Social Progress Research Institute, noting that 'Japan is in a position to have a significant influence on the resolution of global problems and must look beyond its national interests to contribute to the world community'. The heart of the institute is a Policy Forum that its chairman, Takashi Mukaibo, calls a Japanese version of the Club of Rome. Committees will study such issues as global warming, the role of technology, international economic and social systems, and the potential for 'new and constructive relationships among industry, the economy, the society, and culture'.

People and organizations change, Bill Ruckelshaus points out, because they derive some benefit or because they incur sanctions for not doing so. After years of having the stick of penalties applied to their companies, some managers are realizing that the carrot is another important tool. At Du Pont, for example, salary review for mid- and senior-level executives now includes an assessment of their environmental record. And an environmental awards programme for employees is being launched in 1990.

In his first speech after taking over as Du Pont's CEO, E. S. Woolard—one of the new generation of leaders—declared in May 1989 that he intended to make the firm 'one of the world's most environmentally sound manufacturing companies'. Pointing out that environmentalism is the mainstream, he noted that the general public is now the most powerful environmental group in every industrial society.

The head of Du Pont advocates a new ethic—corporate environmentalism—that will guide industry's future actions, and announced a corporate agenda for environmental leadership. In a December 1989 speech, he added that 'industry in the US and around the world has an ethical responsibility to begin to think in terms of sustainable development'. Does this mean he agrees with the Brundtland Commission's definition of this term, which 'requires the promotion of values that encourage consumption standards that are within the bounds of the ecologically possible and to which all can reasonably aspire'? It is too soon to tell.

Mr Woolard ended his thought-provoking first speech as CEO of the largest US chemical company by acknowledging that 'industry has a checkered past of successes and failures in environmental matters, and as a result, manufacturers have been painted many colors in recent years. That will have to change. In the future we will have to be seen as all one color. And that color had better be green.'

* * *

Several recent national initiatives have dealt with important components of a strategy for sustainable industrial development. Canadians have been particularly active. In August 1989 the Canadian Chamber of Commerce published *A Healthy Environment for a Healthy Economy: A New Agenda for Business*. This report from the Chamber's Task Force on the Environment was inspired in part by the Brundtland Commission, writes former Minister of the Environment Tom McMillan in the preface.

In addition to the goal of raising awareness within the Canadian business community about the opportunities and challenges of sustainable development, the report offers practical recommendations for businesses to follow in day-to-day operations: Capture and use what were once called wastes. Use existing recycling programmes or start one locally. Examine energy consumption, waste-handling practices, and transportation systems. Create an environmental-accountability system for owners, shareholders, and employees.

The task force noted that formal statements on the environment have been adopted by several key Canadian industrial associations, such as those on mining, petroleum, and pulp and paper. Of course, anyone can issue a statement of guiding principles. And many more companies will do so as they try to become 'green'. To be effective, these must be backed up by detailed operational guidelines for implementation and a clear corporate method of enforcement. They should be a combination of 'thou shalt not' (pollute, waste energy, and so on) and 'thou shall' (provide employees and the public with information, or consider the environment in all aspects of business planning).

The Canadian Chemical Producers Association, for instance, has both a list of seven guiding principles and codes of practice related to various parts of any chemical business: community awareness and emergency response, transportation, hazardous waste management, and manufacturing. Since 1984, signing the principles and agreeing to adhere to the codes has been a condition of membership in this trade association. The Canadian initiative has proved so successful that associations of chemical manufacturers in Australia, New Zealand, the United Kingdom, and the United States are using it as the model for codes they are currently developing.

To aid in standardizing similar codes and statements on the environment, the Chamber of Commerce in Canada was urged by its task force to develop a Code of Environmental Ethics, covering a dozen points ranging from choosing energy-efficient technologies to providing consumers with information on the environmental impact of all products.

Meanwhile, a key group of Canadian industrialists is working on its own suggestions for such a code. The Business Council on National Issues consists of the CEOs of 150 leading enterprises in the country, covering firms that employ a total of 1.5 million Canadians. In May 1989 BCNI set up a Task Force on the Environment and the Economy; it included such industry giants as Canadian Pacific Forest Products, Consumers' Gas Company, Dofasco, Inco, ITT Canada, and Quaker Oats of Canada.

Soon after announcing formation of the task force, BCNI President and CEO Thomas d'Aquino gave a major address in which he noted that 'reversing the deterioration of the environment on a global basis is the most important challenge facing Canadians and citizens of the world'. The task force's Statement of Principles for Canadian

Business, due out in mid-1990, will try to help companies respond to that challenge.

In the United States, the equivalent of BCNI's proposed code was published in September 1989 by a coalition of environmental organizations and social-investment funds (see Box 5–1). The group included New York City's Comptroller and the Controller of California, individuals who influence a sizeable chunk of investment through their pension funds, so the coalition knew they would get industry's attention.

The code was drafted in the spirit of the Sullivan Principles, created in 1977 by the Reverend Leon Sullivan to guide corporate conduct in South Africa. Unlike those, however, the drafting of the Valdez Principles—named to commemorate the massive oil spill in Alaska's Prince William Sound in March 1989—did not include any industry representatives. The code was launched with much fanfare by the new Coalition for Environmentally Responsible Economies, which claimed it hoped this would be the start of a permanent process in which constituents from all segments of society would collaborate on practical guide-lines for corporations. The coalition expects to announce the first companies that have endorsed the Valdez Principles as soon as operating guide-lines are established.

These contrasting approaches to corporate environmental codes provide a valuable lesson for other national efforts. In Canada's case, the drafting group consists of industrialists, though suggestions for the code are being circulated to a few environmentalists for comments. In the United States, the code was developed by environmentalists and some individuals involved with, but not part of, the corporate world; only after it was held up as a standard were companies invited to help refine it. 'It's sort of come out of the blue', a representative of H. P. Fuller, a chemical manufacturer widely praised for its environmental record, told the *New York Times*.

A better strategy in the future, a strategy that will increase the chance of developing broadly acceptable standards of industrial conduct, is to include from the start all who have a stake in such codes—corporations, environmental and citizens' groups, and government representatives.

*   *   *

Box 5–1. The Valdez Principles

In September 1989 the Coalition for Environmentally Responsible Economies set forth ten broad standards for evaluating corporate activities that directly or indirectly affect the biosphere. The coalition, consisting of representatives of the social-investment and environmental communities, published the following principles to help investors make informed decisions and in the hope of working with companies to create a voluntary mechanism of self-governance:

1. Protection of the Biosphere

We will minimize and strive to eliminate the release of any pollutant that may cause environmental damage to air, water, or earth or its inhabitants. We will safeguard habitats in rivers, lakes, wetlands, coastal zones and oceans and will minimize contributing to global warming, depletion of the ozone layer, acid rain or smog.

2. Sustainable Use of Natural Resources

We will make sustainable use of renewable natural resources, such as water, soils and forests. We will conserve nonrenewable natural resources through efficient use and careful planning. We will protect wildlife habitat, open spaces and wilderness, while preserving biodiversity.

3. Reduction and Disposal of Waste

We will minimize the creation of waste, especially hazardous waste, and wherever possible recycle materials. We will dispose of all wastes through safe and responsible methods.

4. Wise Use of Energy

We will make every effort to use environmentally safe and sustainable energy sources to meet our needs. We will invest in improved energy efficiency and conservation in our operations. We will maximize the energy efficiency of products we produce or sell.

5. Risk Reduction

We will minimize the environmental, health and safety risks to our employees and the communities in which we operate by employing safe technologies and operating procedures and by being constantly prepared for emergencies.

6. Marketing of Safe Products and Services

We will sell products or services that minimize adverse environmental impacts and that are safe as consumers commonly use them. We will inform consumers of the environmental impacts of our products or services.

---

7. Damage Compensation

We will take responsibility for any harm we cause to the environment by making every effort to fully restore the environment and to compensate those persons who are adversely affected.

8. Disclosure

We will disclose to our employees and to the public incidents relating to our operations that cause environmental harm or pose health or safety hazards. We will disclose potential environmental, health or safety hazards posed by our operations, and we will not take any action against employees who report any condition that creates a danger to the environment or poses health and safety hazards.

9. Environmental Directors and Managers

At least one member of the Board of Directors will be a person qualified to represent environmental interests. We will commit management resources to implement these Principles, including the funding of an office of vice president for environmental affairs or an equivalent executive position, reporting directly to the CEO, to monitor and report upon our implementation efforts.

10. Assessment and Annual Audit

We will conduct and make public an annual self-evaluation of our progress in implementing these Principles and in complying with all applicable laws and regulations throughout our worldwide operations. We will work toward the timely creation of independent environmental audit procedures which we will complete annually and make available to the public.

Source: CERES Project, the Social Investment Forum, Boston, Mass.

International organizations representing the business community have also been grappling with the need for sustainable industrial development. Even before the Brundtland Commission's report was published in 1987, the environment had moved up on the agenda of these groups. In 1984 some 500 industrialists, environmentalists, and government representatives from seventy-two nations gathered in France, at Versailles, for what the organizers called a milestone event: the World Industry Conference on Environmental Management (WICEM).

Those attending heard Prime Minister Fabius of France open the meeting by saying that he doubted if such a gathering could have occurred twenty or even ten years before. The new attitude evident in Versailles was summarized by John Elkington and Tom Burke in

The Green Capitalists: 'Increasingly, environmentalists are not asking industrialists to put themselves out of business, but to recognize that the worldwide drive for sustainable development will mean new economic opportunities for those companies which are quickest on their feet.'

The meeting was sponsored by the UN Environment Programme and the International Chamber of Commerce, the leading world business organization. ICC first published environmental guide-lines for world industry in 1974, and has an active Commission on Environment. The interest its broader membership had in these global issues, as evidenced by attendance at WICEM, led ICC in 1986 to set up the International Environmental Bureau (IEB). This non-profit association of business firms is a specialized division within ICC that acts as a clearing-house for information on successful business technologies for dealing with environmental problems.

Within months of the publication of *Our Common Future*, the IEB held a meeting of key members to assess the impact of the report and discuss how industry might help shape the international environmental agenda in the future. While expressing reservations about some specific recommendations of the report, the meeting endorsed the central concept of sustainable development and called on industry to contribute to the process of addressing the issues raised by the Brundtland Commission.

ICC's own contribution was furthered in late 1989 by a new brochure on sustainable development, aimed at helping business executives increasingly drawn into debates on the subject. ICC President Peter Wallenberg summarized the group's commitment to change: 'The onus of proving that sustainable development is feasible rests primarily on the private business sector. . . . We should seize the business opportunities offered by green consumerism, recycling, waste minimization and energy efficiency, and at the same time show corporate responsibility and commitment of a high order in reducing the strain on the environment and in developing innovative solutions.'

The International Chamber of Commerce was also active in preparations for the May 1990 regional follow-up conference in Bergen (see Chapter 8), and contributed to the Warsaw preparatory workshop on sustainable industrial activities held in November 1989. This workshop recommended, among other things, that industry make structural changes that will reduce material and energy use and transportation needs, that companies use the 'cradle-to-grave' approach when considering the impact of their products, and that governments work with industry to develop and evaluate mechanisms to help consumers choose environmentally sound goods.

The organization's participation in the global debate on sustainable development is assured by a WICEM follow-up conference to be held in Rotterdam in April 1991. This will permit both a reflection on relevant recommendations from the Bergen gathering and the preparation of industry input for the UN Conference on Environment and Development in Brazil in 1992.

As *Our Common Future* pointed out, 'transnational corporations have a special responsibility. They are repositories of scarce technical skills, and they should adopt the highest safety and health protection standards practicable.' The Commission also recommended that the codes of conduct for transnationals under discussion at the United Nations should deal explicitly with environmental matters and the objective of sustainable development.

In response to that challenge, the UN Centre on Transnational Corporations developed 'Criteria for Sustainable Development Management of Transnational Corporations', in consultation with governments, environmental groups, and universities, in addition to major multinationals. The final draft, considered in April 1990, laid out ten 'first corporate steps towards an environmentally sustainable direction'. Among them were recommendations to reward employees who come up with new environmentally sound products and processes, to do environmental audits of ongoing activities and sustainable-development assessments of major investment and operating decisions in the future, and to announce significant efforts to reduce use of natural resources and minimize the generation of wastes.

*   *   *

Although the international trade union movement has had no equivalent of WICEM, both the International Confederation of Free Trade Unions and the World Federation of Trade Unions (WFTU) have issued statements on sustainable industrial development. And in April 1989, WFTU—whose members represent 214 million workers—announced it had set up a Trade Union Commission for the Environment, based in Prague, to co-ordinate and promote the activities of its members. The Commission announced that it reflects the concern of workers and people all over the world with regard to environmental damage, the increasing imbalance between industrial activities and living conditions, and the worsening of working and health conditions in factories.

In the United Kingdom, the General Council of the Trades Union Congress (the collective voice of British unions) presented a statement entitled 'Towards a Charter for the Environment' to its 1989 congress. This pulled together work already done on these issues and announced the creation of a special Environment Action Group to co-ordinate union activities at all levels.

Not surprisingly, unions' concerns about occupational health issues have often been the entry point for their contribution on environmental issues. The connection between employees' health and such business policies as accident preparedness and the use of chemicals like pesticides in the workplace is clear. As the WFTU General Council pointed out at its October 1988 meeting, 'neither accidents nor pollution are inevitable; they can be foreseen and avoided by taking measures in factories and workplaces'.

The International Federation of Plantation, Agricultural and Allied Workers (IFPAAW), to cite but one example, published a book on the hazards of pesticides that is part of an audio-visual training package used around the world. In an interview about the impact of the Brundtland Report on the trade-union movement, IFPAAW General Secretary Borje Svensson said the members' first concern was with people's health, but that the connection of that to the health of the land is increasingly being made. As a result, the organization's recent action plan on reducing the use of pesticides includes saving the global environment as one of its goals, along with making the working environment safe for workers in agriculture and related sectors.

At the national level, union leaders are even starting to raise environmental issues in their collective-bargaining sessions. In Louisiana, for instance, in the American South, the Oil, Chemical and Atomic Workers Union successfully negotiated a new contract with the major chemical producer BASF in December 1989 largely because of a public campaign about the company's environmental record. The State Attorney General's office told the *New York Times* that over five years the union had raised the level of awareness about environmental issues in Louisiana and helped persuade the government to regulate industry more closely.

* * *

All the initiatives discussed so far have not arisen solely from the corporate world's new-found understanding of sustainable development. As ICC President Wallenberg's comment indicates, companies are to a certain extent reacting to the rise of 'green consumerism'. A new breed of buyers concerned about the environmental impact of their purchases is particularly evident in the United Kingdom.

The UK Department of Trade and Industry, in advice on 'Protecting the Environment' from its newly formed Business and the Environment Unit, notes that consumer demand may soon be a greater impetus to environmental standards than regulations are. Indeed, in late 1988 *The Green Consumer Guide* by John Elkington and Julia Hailes shot to the top of the UK trade paperback best-seller list. Within a year of publication, 250,000 copies had been sold.

By April 1989, Mintel research reported that 27 per cent of the British could be called green consumers. An impressive 75 to 80 per cent claim to know about CFCs in aerosols, acid rain, leaded petrol, and nuclear-waste issues. Soon after, MORI (Marketing and Opinion Research International) reported on an astounding change it had tracked over the year: the share of the general public who selected one product over another because of 'environment-friendly' packaging, formulation, or advertising leapt from 19 to 42 per cent—representing sixteen million adults.

Similar strong feelings for the impact of their purchases have been noted among consumers on the other side of the Atlantic. An opinion poll by the Michael Peters Group found that 89 per cent of Americans are concerned about the impact their purchases have on the planet, that 78 per cent are willing to pay extra for products packaged with recyclable materials, and that 53 per cent had refused to buy a product during the preceding year because they were worried about the environmental impact of the item or its packaging. The US version of *The Green Consumer Guide*, written by Joel Makower and published in April 1990, appeared to have a ready audience.

And in Canada, research done for the supermarket-chain giant Loblaw as they planned to launch a green product campaign found that half of all consumers would switch to a store selling merchandise they believe to be environmentally safe. Earlier opinion polls had indicated that 80 per cent of consumers would pay 10 per cent more for products that have a low environmental impact. The Chamber of Commerce highlights one of the many reports on public opinion: 86 per cent are willing for their 'green conscience' to cost them $100 more in expenses per month.

Canadian stores may soon feel the effects of this sentiment: *The Canadian Green Consumer Guide* by the environmental group Pollution Probe, their version of the Elkington and Hailes book, has given consumers the information they need to buy green. And in a much livelier format—chapters on the home, gardening, transportation, waste management, and so on are full of colourful cartoons and clever illustrations to deliver serious messages about how to change buying and consumption patterns. Within two months of its November 1989 launch, 92,000 copies had been sold, a strong indication of the desire of Canadian consumers to be 'green'.

While the emphasis on a purchase's impact varies a little among countries, reflecting priorities in different cultures, in general the advice given to green consumers follows these guide-lines developed by SustainAbility, Ltd., of London. The organization recommends that consumers avoid products that are likely to:

- endanger the health of the consumer or others;
- cause significant damage to the environment during manufacture, use, or disposal;
- consume a disproportionate amount of energy during manufacture, use, or disposal;
- cause unnecessary waste, either because of overpackaging or because of an unduly short useful life;
- use materials derived from threatened species or from threatened environments;
- involve the unnecessary use of or cruelty to animals, whether this be for toxicity testing or for other purposes; or
- adversely affect other countries, particularly in the Third World.

SustainAbility is a small environmental-policy and corporate-communications company set up John Elkington and Tom Burke to promote environmentally friendly economic growth. Formed in 1987, it was quickly thrust on to centre stage in the flurry of companies anxious to court the new market.

One of its first products was *Green Pages: The Business of Saving the World*, a comprehensive and most useful compendium of environmental issues and who was doing what about them. One of the strengths of SustainAbility is indicated by this book's inclusion of contributions from Petra Kelly, a well-known Green party member of the West German Bundestag, as well as Christopher Harding, Chairman of British Nuclear Fuels (BNF). Indeed, *Green Pages* has both a full-page advert from this nuclear power company and the editors' report that BNF is the firm chosen by environmental groups as having the worst record on these issues. SustainAbility clearly acts on its belief that all views are worth airing.

The success of the group's next project, *The Green Consumer Guide*, surprised even them, John Elkington has noted; by 1990, versions of the guide were available in ten other countries. In the United Kingdom, this was soon followed by more specific SustainAbility guides, on supermarket shopping, for example, and office supplies.

Some manufacturers have been quick to see the potential of this new market. Tesco, one of the largest British food chains, introduced 'Tesco Cares' labels on its own brands in January 1989, and claims to be the nation's Greener Grocer. Sainsbury's soon followed suit, dubbing itself the Greenest Grocer. In Canada, Loblaw introduced more than 100 green products in June 1989 after the marketing director returned from a trip to Britain where he saw the success of the new campaigns. Loblaw has been taken to task by environmental groups, however, for not releasing the criteria it uses to include products in the new line.

The third largest retailer in the United States recently joined the rush to promote green-ness. In August 1989, Wal-Mart Stores, Inc., challenged its 7,000 suppliers to reduce the environmental impact of the manufacture and packaging of their products. Special signs will alert Wal-Mart customers to products that companies have improved. A mailing to all suppliers about this new programme announced that 'by sharing with our customers Wal-Mart's commitment to our land, air and water . . . we're convinced that we can begin to make a difference!' As Wal-mart has 1,326 discount retail outlets across the country and $25 billion in annual sales, manufacturers are likely to pay close attention to its challenge.

*   *   *

With so many people landing on the green-product bandwagon at once, it stands a good chance of tipping over. Consumers who think they are taking their own small steps to save the earth and who later find they were the victims of a marketing ploy could get turned off the whole idea. One watchdog hoping to avoid this is Friends of the Earth in Britain.

In December 1989 this group issued Green Con Awards, giving black marks to companies for 'factual inaccuracy, significant omission, and exploitation of the public's fears and ignorance of some environmental issues'. The winner of their main award was British Nuclear Fuels: the company's ads equate being green with being free from carbon dioxide emissions, conveniently not mentioning several unsolved environmental problems associated with nuclear power, like disposing of the radioactive wastes. Friends of the Earth warns that 'green hype can be no substitute for genuine change'.

Another safeguard against consumers being conned is the development of official labelling programmes. These have been going since 1978 in West Germany, were launched recently in Canada, Japan, Norway, and Sweden, and are being considered in Britain, Denmark, France, the Netherlands, and the United States. Most look to the West German system as a model, although a new trend is to consider products that minimize the resources used in manufacture and marketing, as well as the impact of disposing of items—the cradle-to-grave approach.

West Germany's Blue Angel system, recognized by 80 per cent of West German consumers, over a decade approved some 3,000 products in fifty categories, from low-pollution paints to wallpapers made from recycled paper, with most of the angels being affixed to products since 1986. The Ministry of Environment has reported that the resulting changes in purchases led to a notable reduction in the use of carcinogenic substances and CFCs. A 1989 report on eco-labels by Environmental Data Services Ltd. of London concluded that 'the Blue Angel scheme has undoubtedly been successful in stimulating consumer awareness and industrial innovation, and in reducing the environmental burdens created by consumer products'.

The Canadian labelling programme, launched by Environment Canada in October 1988, is called Environmental Choice. Producers pay an annual fee of up to Can.$5,000 (about $4,200) once they are accepted in the programme, and products are screened by a panel representing environmental, business, and consumer groups.

The first three categories Environment Canada accepted applications in were re-refined oil, construction material produced from wood-based cellulose fibres, and plastic products made from recycled plastic. In February 1990 cloth diapers, newsprint and fine paper from recycled stock, and water-based paints were among the categories added to the list.

Approved products are able to use the Choice symbol—a maple leaf formed by three doves, representing consumers, industry, and government. Corporate interest in participating in this programme will no doubt be aided by the government's stated intention to buy Choice-labelled products whenever possible. This goes hand-in-hand with the government's decision to become, as Environment Minister Lucien Bouchard put it, 'a model environmental institution . . . providing national leadership in our purchasing policies, recycling, and the simple reduction of waste'.

Japan's label is called an eco-mark, and uses the symbol of the earth embraced by two arms, to symbolize 'saving the earth with our own hands'. The logo will be used not only on products but also as part of local recycling campaigns and other efforts to encourage consumers to contribute to environmental protection. It was developed by the Japanese Environment Agency, and covered forty-four products when launched in February 1989.

The first list in Japan included such items as aerosol deodorants without CFCs (as aerosols with CFCs are still available there, unlike in the United States) and books and magazines printed on recycled paper. The list of products they hope will qualify for the eco-mark includes electric cars and solar-powered water heaters, reflecting the Environment Agency's hope that the scheme will spark innovative industrial development.

Similar programmes are being developed in the United States and Britain, though on a different basis. The US effort is a private one, being put together by a new non-profit group called the Alliance for Social Responsibility. Initial plans for a Seal of Social Responsibility were dropped as being too broad, and attention focused instead on developing criteria for a Green Seal awarded to a few products in time for Earth Day 1990.

The Alliance works closely with the Council on Economic Priorities in New York, publishers of *Shopping for a Better World*, a small booklet that rates 1,300 companies and specific products in terms of the producer's record on advancement for minorities and women, use of animal testing, investment in South Africa, positive steps on the environment, and similar criteria. This pocket-sized guide was a runaway hit in the United States, with 400,000 copies sold in a year.

In the United Kingdom, a government-run programme is being considered, more akin to the Canadian Choice label. During 1990 a Steering Committee is expected to consider the criteria and operating procedures for such a programme, as well as the advisability of waiting for a labelling scheme to be introduced for the whole European Community. At the moment, it appears that a cradle-to-grave approach will be favoured, rather than just letting consumers know the environmental impact of packaging and disposal of their purchases.

For all eco-label programmes, the criteria used to identify products are obviously crucial. For a while, the Blue Angel system was criticized by some for not keeping up with technological developments that allowed better environmental performance, and for not assessing overall product quality. To be credible as well as useful to consumers, labelling programmes must publicize the criteria for their choices and include consumer- and environmental-group representatives on their selection panels.

*   *   *

Some businesspeople are responding not only to the push of green consumers but to the pull of their own convictions. For them, protecting the environment is not a tactic, but part of an overall business strategy. The distinction is an important one, Alan Gussow, Chairman of the Board of Friends of the Earth in the United States, points out. Companies always searching for the best tactic to lure customers will move on if they think the environment is no longer 'flavour of the month'. Companies that are truly green will realize that minimizing the impact of the production and marketing of their products is a sound long-term business plan.

In Britain, the best known of this new breed is Anita Roddick, founder and owner of The Body Shop. This chain of cosmetic stores has grown from her initial investment of about £7,400 ($12,500) in a Brighton shop in 1976 to some 450 stores in thirty-four countries, grossing some £74 million ($125 million) annually. Cosmetics are sold in reusable containers, so customers can get refills, and all paper goods are on recycled stock. As the UK shops alone use 20 million paper bags a month, all with environmental messages stamped on them, the market created for recycling mills is notable.

Although Anita Roddick says she did not set out to 'capture' green consumers, because they did not exist when she opened the first Body Shop, she now boldly claims that the stores are a force for social change. She uses them to campaign for numerous environmental causes, from CFCs in aerosol cans to rain-forest destruction. Over three weeks in 1989, a half-million signatures were collected on petitions to the Brazilian government to protect the forests; these were eventually dumped on the front door of the embassy in London, with appropriate media coverage.

The Body Shop pays First World prices to Third World suppliers, and has helped to start a number of companies in developing countries when Anita Roddick has identified products that The Body Shop can stock. If other firms want to buy goods from these fledgling companies, they must agree to a charter of ethical behaviour that Ms Roddick developed to protect the rights of the Third World workers, mostly women.

In the United States, Anita Roddick would be called an eco-entrepreneur. Probably the most famous American versions of this new type of businessperson are Ben Cohen and Jerry Greenfield, ice-cream makers extraordinaire and promoters of what they call progressive capitalism. As they prepare such flavours as Cherry Garcia and Dastardly Mash, staff in the Vermont headquarters of Ben & Jerry's look like they have never left the sixties. But they are really a vision of the smart business of the nineties.

Each year the company gives 7.5 per cent of its profits to charity, from national rain-forest groups to local peace coalitions. The newest Ben & Jerry's ice cream is directly related to the founders' environmental concerns: Rain Forest Crunch uses cashew and brazil nuts purchased from indigenous groups in Brazil, in a bid to show the economic viability of keeping forests intact.

Necessity has literally mothered invention in the case of several other eco-entrepreneurs. In the mid-eighties, new parents Jenny Gilles and Cathy Trojanoski in British Columbia decided not to add to the nearly 20 billion disposable diapers piling up in North American landfills each year. So they came up with a no-pin, Velcro-fastened cloth diaper, and named their new company Babykin's Products of Canada Ltd. By the autumn of 1989 they were churning out 60,000 diapers a month and looking to distribute in the United States as well.

Another new mother, Sarah Redfield of Maine in the United States, was appalled to discover in 1986 that she could not find organic apple juice for her son. As she had been the Associate Commissioner of the state's Department of Agriculture, and had served on the Pesticide Control Board, she was well informed about the growing concern over the pesticide used on apples, Alar. So she teamed up with a friend—a friend who happened to be the former Commissioner of Agriculture for Maine—and started Simply Pure Foods in her basement. Three years later, she had a full-scale production facility and thirty staff.

'I'm just a mother who got mad when she could buy organic pet food but not organic baby food', Sarah Redfield drily explains. It helped, no doubt, that the founders were well known in the state and in the farming community, and were so knowledgeable about what qualifies as organic agriculture. But the lesson Ms Redfield draws from her experience is an important one: 'When you have a good idea, don't let experts discourage you. Do it. Most experts are living in a world that is no longer real for most people.'

Some environmentalists—occasionally called deep ecologists or dark Greens—are uneasy with the whole concept of the green consumer, as it implicitly promotes the very thing the world needs to reduce: consumption in the industrial countries. But when their alternative is to tell people they must stop shopping altogether, they risk losing their audience.

People in industrial countries can meet their food, shelter, clothing, and other needs in a 'greener' manner. At the same time, conspicuous consumption—acquiring more things just for the sake of 'owning' them—must be reduced, and then eliminated. The more information consumers have about the impact of their purchases, the sooner they will start considering whether they really need each item.

Jonathon Porritt, former head of Friends of the Earth in Britain, is clear on the proper context of this new phenomenon:

> Green consumerism . . . can only be useful in a political sense if it is part of a transitional strategy. And the problem with any transitional strategy is its potential to lull people into the belief that the transitional stage is actually the goal. Those who promote the benefits of Green consumerism today may themselves see it as a first tiny, faltering step

on the road to our Green utopia, but others may be more inclined to see it as the end of the road.

That road does not end until purchasers, and producers, stop consuming the capital of the earth, and live instead on the interest provided by its resources.

*   *   *

This chapter has dealt nearly exclusively with signs of hope in industrial countries. But this should not distract attention from the pressing need to help developing countries, as well as East European ones now rushing to embrace western technologies, avoid making the very mistakes others are now trying to correct. Chapter 7 considers ways to finance alternative technologies, and what some industries in developing countries might consider producing.

'Of critical importance is technology transfer, so that environmentally benign technologies can be widely used in developing countries', noted Dr Mostafa Tolba, UNEP Executive Director, in a letter with comments for this book. 'The industrialized countries recognise they have made a mistake. The South cannot follow such mistakes, by using ozone-destroying CFCs, halons, carbon-intensive fossil fuels that contribute to global warming. Nor can the South be penalized by pursuing environmental protection.'

For some in the Third World, concern about the environment focuses on outrage at the thought of being a dumping ground for industrial nations' debris—toxic debris, at that. (See Chapter 2 on international efforts to stop this practice.) Beyond that, green consumers may seem irrelevant in developing countries, one more luxury they cannot afford. Yet if consumers, and investors, and auditing groups in the West follow their environmental concerns to their logical conclusions, the practices of multinationals in the Third World can also be affected.

Consumers need to start looking backwards as well as forwards, as Alan Gussow puts it. They should certainly be concerned about recycling paper bags, soda bottles, and the CFCs in their car air-conditioners. But they also need to find out the environmental impact of the processes that created the products they buy. Companies that do cradle-to-grave audits can provide consumers with these details. Applying an even broader criteria—a sustainable-development look backwards—consumers need to think about whether the living conditions of people paid a minimum wage to produce the basic materials leave those individuals no choice but to degrade their local environment.

Product labels that contained all this information would no doubt overwhelm shoppers. The envelope of a piece of direct mail received recently in Washington was already suffering from politically correct information-overload: 'printed on 100% recycled, unbleached paper with a bio-degradable (non-plastic) window.' Still, getting as much information as possible to consumers should be the goal of watchdog groups, of governments, and of companies that want their business. One of the points of any labelling scheme is to make consumers stop and think about their purchases; applying a sustainable development criteria should make them stop and think whether they really need the item at all. People in the nineties who carefully consider the impact of their purchases on all the components of sustainable development will have to be called more than green — perhaps deep green consumers.

Other developments that can be expected during this decade include an increasing sophistication on the part of business on how to meet this criteria of being 'green'. Pollution control is big business in western industrial countries and that will continue, for there is a great deal of pollution to clean up.

But the 'upstream' side of sustainable development will also become big business in the nineties. As Worldwatch Institute points out in a recent attempt to picture a sustainable society, the world is going to need more wind prospectors, solar architects, specialists in biological pest-control methods, and so on: 'Many people will find their skills valued in new or expanded lines of work. Petroleum geologists may be retrained as geothermal geologists, for example. . . . Since planned obsolescence will itself be obsolete in a sustainable society, a far greater share of workers will be employed in repair, maintenance, and recycling activities than in the extraction of virgin materials and production of new goods.'

The contrast of these opposite ends of the green business spectrum was evident in two meetings held in March 1990. GLOBE '90, in Vancouver, was put together by a unique combination of government and the private sector; the organizers now hope to hold meetings every two years on the theme of 'global opportunities for business and the environment'. Some 17,000 participants attended this first conference and trade fair; companies represented on the exhibition site leaned heavily towards the clean-up end of business—

firms such as Browning-Ferris Industries, Laidlaw Waste Systems, Filtration System Products, and Wastewater Treatment Systems.

A few weeks before, some 150 people attended a conference in California on 'Environmental Entrepreneuring: The New Reality for America's Small Business Movement'. Participants heard about advertising green products, direct marketing to eco-consumers, packaging design to minimize waste, and case studies on the renewable energy, organic food, and recycling and composting industries. This gathering was the second on this theme organized by *In Business*, a bimonthly published in Pennsylvania that has declared itself to be the magazine for environmental entrepreneuring.

As the imperative of balancing the environment with economic development becomes clearer—and as opportunities for large and small companies that embrace this as a strategy, not a tactic, become evident—conferences and trade fairs will feature the Anita Roddicks and Sarah Redfields of the world right alongside the sales staff of Browning-Ferris and Wastewater Treatment Systems.

6

Delivering the Message

The environment is big news—seen on television daily, discussed on the radio, featured on front pages the world over. The media seem to have 'discovered' the environment, and in so doing are finally starting to ask how to cover this 'new' story. The answers they come up with will determine what information reaches viewers, listeners, and readers in the nineties—a decade being called crucial because of the important decisions that must be made if the world is to move towards sustainable development. Pressure to make the right decisions, and support for the politicians that make them, must come from a public well informed about complex issues.

One Saturday in September 1989, some 100 journalists, news-paper editors, television reporters, and environmentalists gathered in Washington, DC, at the Smithsonian Institution to hear twelve leading US scientists answer the question 'The Global Environment: Are We Overreacting?' That very same day, some 250 journalists, environmentalists, and peace activists met outside Stockholm for an international conference on the media and our common future. Five thousand miles apart, key segments of this influential industry thought about how to cover what may be the biggest story of their lifetimes: our survival.

Meetings such as these highlight the important role of the media in the time of rapid change the world is facing. The ripple effect of reaching journalists is enormous. The Smithsonian workshop, for example, converted at least one sceptic. The following Monday David Gergen, editor-at-large at *U.S. News and World Report* and a well-known conservative, told the one million listeners of 'All Things Considered' on National Public Radio about his amazement at the environmental horrors predicted, and about the depressing gap between the scientists at the meeting and the politicians who run Washington. 'Official Washington', he concluded, 'needs to sit up and pay sharp attention. . . . And Congress and the President need to think and act much more boldly about the environmental challenge.' Hearing Ronald Reagan's former Director of Communications be so convinced, and so convincing, about the sad state of the environment was quite a pay-off for a one-day conference.

The workshop was also attended by Ben Bradlee, editor of the *Washington Post* and therefore one of the most powerful individuals in the US media today. His lack of interest in the environment has been renowned, and he now tells the story himself of how a few years ago a reporter tried to get him to run a front-page story on the ozone problem. At the editorial meeting, Bradlee mimed the squirting of an underarm deodorant, and laughed. The connection between aerosol cans and the depletion of the ozone layer was to him far-fetched.

Then in 1988 Bradlee went down to the Amazon on a tour arranged by Tom Lovejoy of the Smithsonian, a leading expert on the threats to species and ecosystems around the world. Another convert was born; now the *Post* has three environmental reporters, and stories appear on the front page more often. As Jon Tinker, Director of the Panos Institute, notes, 'ignorance is a major contributory cause of environmental degradation; awareness is the first step towards environmental reconstruction'. That applies equally to editors behind a desk in Washington and radio listeners wielding a machete in the rain forest.

*   *   *

By 1989 there was a veritable explosion of media interest in anything to do with the environment, egged on by the oil spill in Alaska, the outcome of the Group of Seven summit in Paris that July, and the rise of green parties in Europe. Reporters and editors, in their usual role as mirrors of public interest and opinion, had started to improve their coverage of environmental issues a few years ago. But by 1989 the gravity of environmental problems really began to sink in.

The *New York Times* provides an example of this shift. Environment reporter Phil Shabecoff noted in November 1989 that the first story he did on the greenhouse effect ran in 1979. 'It said many of the same things about what was likely to happen as last year's stories did. It ran on page 42. I guess you could say it took nine years for the greenhouse story to graduate to page 1.'

So the environment is no longer a dead-end beat on papers, something young reporters get stuck with. Will this new interest fade? Frances Cairncross, environment editor at *The Economist*, doesn't think so: 'In the past year, many people have thought seriously for the first time about the environment. . . . This new public interest in the environment may prove more durable than the burst which accompanied the Stockholm conference, nearly twenty years ago. This time, the emphasis has been on the extent to which good economics and sound environmental policies are linked.'

The next decade is bound to bring even more special magazine issues, television series, and in-depth newspaper reporting on these problems. At the moment, the best example of the new trend is the 2 January 1989 edition of *Time*. Replacing the magazine's normal Man or Woman of the Year cover story, the editors chose Endangered Earth as Planet of the Year, and devoted thirty-three pages to 'the looming ecological crisis'. The issue generated more letters to the editor than any previous cover story.

It was not the first time the editors had chosen a non-human for their annual special issue: in 1982 the computer was picked as the Machine of the Year. But it *was* the first time they had done more than just report on who, or what, their choice was. While thoroughly answering the question in the cover story—'What on Earth Are We Doing?'—they included recommendations on what nations should do about the loss of biodiversity, global warming, hazardous wastes, overflowing landfills, and population growth.

In an even more pointed section, *Time* said the United States 'must be in the vanguard of the effort to solve the earth's environmental crisis', and listed eight steps the government should take unilaterally and immediately. Their mid-December 1989 'Endangered Earth Update' took the same tack, including again boxes with recommendations on 'what can be done'.

Newsweek, *Time*'s weekly competition, soon weighed in with its own cover story, right after the 'Green summit' in Paris. Under the title 'Cleaning Up Our Mess: What Works, What Doesn't and What We Must Do to Reclaim Our Air, Land and Water', contributing editor Gregg Easterbrook looked mainly at US problems in an extended essay about the state of the environment.

Taking a broader view, *The Economist* of London devoted eighteen pages to the environment in a special section on 2 September 1989 called 'Costing the Earth'. Starting with the observation that 'never have so many politicians seized so quickly on one idea', the magazine laid out why sensible green politics are going to require people, governments, and companies to make 'immense changes'. All that is needed, concluded environment editor Cairncross, is the will.

Canada's *Financial Post* devoted the June 1989 issue of *Moneywise*, the magazine that goes to all its subscribers, to investing in the environment. Readers learned about companies that are putting environmental protection into production, about five people who have folded their environmental commitment into their businesses, and about practical steps for 'eco-conscious' home-owners.

One year later, *Business Week* in the United States will have its own say on the subject. Under the title 'Agenda for the 21st Century: Managing Earth's Resources', the special supplement is being put together with an unusual editorial board that is drawn from industry, science, government, and environmental groups. Even more unusual, 10 per cent of the net advertising revenue will be donated to a programme of grants for promising citizens' groups in the Third World.

In one of the most impressive contributions from daily newspapers, the Southam chain in Canada spent a year putting together a special section called *Our Fragile Future*. The twenty-four page supplement distributed in almost two million copies of papers across the country in October 1989 included features on 'Our Love Affair With the Car: The Road to Ruin' and 'Learning to Live With Less'. A poll done for the project found, among other things, that nearly one in three adults now considers the environment the most pressing issue in Canada, compared with one in ten just a year earlier.

In addition to this mainstream and moneymakers' fervour for anything 'green', concern about the environment has reached into the most unlikely places: a *Ms*. piece asserting 'It's Not Nice to Mess with Mother Nature', a *Vanity Fair* article on 'The Hot Issue', a *Rolling Stone* essay on 'Covering the World; Ignoring the Earth', a *Family Circle* special section telling its twenty-two million readers how to 'Save Our Earth', and even a piece in *Vogue* called 'Planet

Stricken' that claims to report from the frontiers of America's newest religion.

But how many special issues will the public take? The more challenging task is introducing continuing coverage of what does not fit into the traditional definition of 'news'. Donatus de Silva of the Panos Institute points out that 'achieving sound development . . . is a process that takes time. Information about development and environment is also a process. You cannot treat an environmental event as a one-shot affair: write one article and forget about it. Deserts do not suddenly spread overnight. They expand as a result of a series of causes over a period of time.' In *Rolling Stone*, Mark Hertsgaard neatly summarizes this dilemma: How do you take a picture of the earth getting hotter?

Time's answer is to have a regular section on the environment. Under the 'Endangered Earth' logo developed for the special issue, the magazine has continued to cover these issues each week whether a breaking news-story needs to be reported or not. In England, every paper has an environment correspondent now, with the *Observer* deserving credit for having Geoffrey Lean in that slot the longest. The *Guardian* introduced an environment section in September 1989 that runs each Friday. In Venezuela, two Caracas dailies, *El Nacional* and *Economia Hoi*, each run an 'Ecologia' page every week. And *El Nacional* has a 'Sciencia Amena' (friendly science) column every day that often deals with environmental issues.

Even more ambitious is a new monthly called *China Environment News*. Published in Chinese and English in Beijing, and sponsored by the Environmental Protection Commission of the State Council, its editors claim it is 'the first national newspaper dealing with environmental protection in the world'. The premier issue in English, in August 1989, included stories on the Montreal Protocol on the ozone layer, schoolchildren who are 'green' guards, increasing damage to forests and pastureland in China, and the Group of Seven summit in Paris. Similar efforts elsewhere could help policy-makers and the public monitor the state of the earth more closely.

Other countries may not have national newspapers devoted to tracking environmental developments, but readers are being swamped by magazines that do so. For years, members of environmental organizations have received beautifully produced, written, and photographed magazines on threats to wildlife and, occasionally, ourselves. And British readers could turn to *The Ecologist*, *Resurgence*, and the *New Internationalist*, though they were unlikely to find them at the corner news-stand. But now anyone browsing at magazine racks has many choices, from publications by clever marketers trying to capitalize on the latest hot topic to glossy attempts by long-time environmentalists trying to reach a broader audience.

January 1988 saw the launch of *World Watch*, a bimonthly from the Washington-based Worldwatch Institute, a well-respected research organization started in 1974 by Lester Brown. After only two years of publication, circulation reached 30,000 and *World Watch* articles were being reprinted or adapted in daily newspapers throughout the United States. By special arrangement with a local environmental group, a Japanese version of *World Watch* is also available.

A few months later, *Buzzworm: The Environmental Journal* appeared, and tried to capture the general reader in the United States. The title is an old western term for rattlesnake, which the editors hope symbolizes their effect on readers: they buzz and you react. It has a heavy focus on adventure travel and pretty pictures of animals, but may help raise awareness of broader environmental issues among readers not normally members of traditional conservation groups.

Two other new US entries are the simply titled *E Magazine* and one with the more clever title of *Garbage*. Both are printed on recycled paper, have lively and inviting graphics, and are packed with information on how to live 'green'. And both are filled with advertisements for recycled notecards, environmental groups seeking new members, money-market funds that invest in companies that are allegedly environmentally friendly, and so on. *Garbage* appeared to be an immediate hit, reaching 100,000 circulation as soon as it appeared. Also expected on the stands in 1990 are new magazines *Planet* and *The New Environmentalist*.

Though it does not fit into the category of a new magazine, *Greenpeace* can be thought of as being reborn. Following the extensive mass-media coverage of the organization's loss of the ship *Rainbow Warrior* to a French bomb, Greenpeace membership skyrocketed (see Chapter 4). As all members receive the magazine, *Greenpeace* claims an astounding readership of 2.4 million in its circulation area of Australia, Canada, and the United States. It may be doing more to deliver the message about the state of the environment than all these other magazines combined.

*   *   *

When it comes to boosting public awareness, television would seem to have an easier job. With a picture being worth a thousand words, producers can certainly raise alarm by showing dying forests in West Germany, a spreading oil-spill in Alaska, children in Ethiopia bloated with malnutrition, and the burning of the rain forest in the Amazon.

They have not hesitated to do so. Soon after the launch of *Our Common Future* in 1987, the BBC ran an eleven-part series called 'Only One Earth', echoing the title of the influential book written by Barbara Ward and Rene Dubos for the 1972 UN Conference on the Human Environment. The series, which covered people finding solutions as well as the problems they face, was also seen in the United States several times on the Public Broadcasting System.

A broader effort to tap the power of television is to occur in May 1990, following the conference on the Brundtland Report held in Bergen, Norway. 'One World' pulled together fifteen European broadcasting organizations last year to develop a week-long package of programmes: a launch film presenting a snapshot of the world's environmental problems, a ninety-minute drama about environmental refugees marching on Europe, a press conference with world leaders, several one-hour programmes about the themes addressed in Bergen, and various musical events.

Soon to follow in the United States will be 'Race to Save the Planet', a ten-part series from WGBH in Boston premiering in October 1990 on many Public Broadcasting System stations. Inspired by Worldwatch Institute's *State of the World* books, the series focuses both on the problems facing policy-makers in the next decade and on constructive ideas and new approaches from all over the world.

Aside from these specials, several organizations aim to improve television coverage of environmental issues. In 1984 Robert Lamb set up Television Trust for the Environment (TVE) in London. This unit has produced documentaries on numerous sustainable-development issues, has worked to build programme-making capacities in developing countries, and since January 1988 has

published an invaluable quarterly guide to development and environment films world-wide. A French edition of its *Moving Pictures
Bulletin* is produced by L'Association des Trois Mondes in Paris.

In the United States, the Better World Society (BWS) has as one
of its goals the production of documentaries on environmental
issues. Since 1985 more than thirty have been released, including
'Island of Peace', about Costa Rica's extensive national parks and
its refusal to have a national militia; 'Jungle Pharmacy', about the
wealth of medical and other scientific knowledge in tropical forests;
and 'Voice of the Amazon', about Chico Mendes's life, and his
death at the hands of assassins because of his efforts to protect the
rain forest. BWS was founded in 1985 by cable-network mogul Ted
Turner, now a formidable force in television around the world and
a recent convert to concern about the environment.

For television, as for the print media, the more challenging task
is to cover the environment day in and day out. All three major
US networks focused on the environment for seven days on their
prime-time evening-news programmes during different weeks in
1989. But by far the most comprehensive effort on television comes
from Ted Turner's Cable News Network (CNN), which has had an
eight-member environmental unit since 1980. CNN has the first,
and still the only, environment editor of a national network—
Barbara Pyle.

Ted Turner's personal conviction about the importance of this
'story', and about the need for viewers to learn more about it,
means CNN runs one to three items a day on the environment. TNT,
the Turner cable network that shows old as well as made-for-cable
movies, has featured such films as an anti-nuclear drama called
Nightbreaker and *Incident at Dark River*, about a man whose daughter
is killed by toxic waste dumped by a local factory. Turner
Broadcasting System, operator of the nation's most watched cable
network, has developed a cartoon series with a hero called
'Captain Planet', whose arch-enemies are global warming, acid rain,
over-population, and the like.

Using a medium with such a broad appeal is not unlike what
Televisia, the most widely watched television network in Mexico,
has done for years. There, characters in a series of popular soap
operas weave mention of family planning and population pressures
into weekly episodes loaded with all the usual relationship problems
found in daytime television. Studies indicate that more than a

half-million women began planning their families for the first time following one year's episodes; contraceptive sales registered an increase of 23 per cent. Televisia is advising networks in other parts of Latin America on their successful formula.

In Brazil, *TV Viva* brings stories about family planning and the hazards of toxic chemicals to Recife's shantytowns on a screen mounted on top of a roving station wagon. Alan Durning reports in *World Watch* that 12,000 people a month are reached by this form of community television, a project of the Luiz Freire Cultural Center. Programmes also deal with such topics as community efforts to build schools, secure rights to their land, and prevent domestic violence. There are many messages to deliver, and many ways to deliver them.

*   *   *

'Hollywood gets ready for the nineties', announced a recent *Vanity Fair* article, 'by turning various shades of green.' Two California-based groups of growing influence, both packed with big-name talent and high-priced producers, have discovered the value of the ripple effect.

The older group is EMA, Environmental Media Association, which bills itself as an entertainment-industry response to the global environmental crisis. Their aim is to 'infuse the popular culture with . . . a climate of concern about our environment and give creative expression to the vision of a healthy future for the planet'. One early indication of EMA's success is the inclusion in the award-winning series 'thirtysomething' of a story about a hazardous waste-disposal facility—not a topic you would expect to build a plot around on prime-time television. Plans are being made to bestow annual 'EMAs' in order to give visibility to exemplary film and television productions on the environment.

The second Hollywood group is a younger, brasher offshoot, formed after some difference of opinion with EMA organizers about orientation and *modus operandi*. Called ECO, for Earth Communications Office, in its first year of existence it dragged some actors and writers out to the Mojave Desert to see the world's largest solar-power facility, and took another group down to Brazil for one of Tom Lovejoy's eye-opening trips up the Amazon. ECO also works with recording artists to include messages about the environment on record jackets, inside compact disks, and as part of audio-cassette packages.

In noting the impact of the ripple effect, Andy Spahn of EMA recounts the story of a seventies situation-comedy star popular with teenagers—the Fonz on 'Happy Days'—who during one episode took out a library card. The next day libraries across the country were swamped by teens applying for cards. As EMA and ECO succeed in including environmental themes in daily television, the same acceptance can occur regarding recycling, riding bicycles, planting trees, eating meat less often, and many other components of a sustainable society.

The film industry is another target of EMA and ECO, and one that has already jumped on the environmental bandwagon. After the tragic assassination of Chico Mendes, for instance, producers and agents rushed down to Brazil to vie for the rights to the life story of this 'ecological martyr'. Mendes was a vocal and articulate representative of the rubber tappers in one part of the Amazon. His efforts to preserve the rain forest from the chainsaws and hatchets of wealthy cattle-ranchers earned him a barrage of bullets as he stepped out of his back door in December 1988. After being courted by everyone from Robert Redford to Ted Turner, Mendes' widow Ilza gave the nod to a Brazilian film-maker, with the revenues from any film to go to a foundation supporting the struggle of rubber tappers to preserve their way of life.

Tales like this are sure to be repeated many times in the years ahead. No less an authority than *The Hollywood Reporter* observes that 'true stories about the world's endangered environment will continue to be "hot" property for film producers'.

The other wing of the entertainment industry—musicians, especially rock ones—has certainly not been immune to the plight of the planet. Since the day 'Live Aid' was seen by a world-wide audience of 1.5 billion, the rock industry has spoken out about environment and development issues. Like the newly converted actors and actresses, pop stars have had to learn a great deal about the environment in a short time in order to give their campaigns credibility. Their enthusiasm has sometimes run ahead of their grasp of complex issues, but in general they have done their homework.

British singer and actor Sting has given a series of concerts in aid of saving the rain forest, started his own foundation towards that end, and toured western nations with a Brazilian Indian, Raoni, at his side to talk about the threats to Indians' livelihoods and very existence. Others similarly involved include the Grateful Dead and Madonna.

Paul McCartney gave a page of his concert booklet over to Friends of the Earth during his late-1989 tour, exposing a whole new audience to one of the first environmental groups to take to the streets. One US rock-video cable channel now airs 30-second 'Earth Alerts', with information on a range of environmental issues and details of how to contact Greenpeace.

In one of the largest group shows of support, some twenty leading rock groups collaborated on a record album called *Breakthrough* that aimed to expose Soviet fans to environmental issues in general and Greenpeace in particular. In the first month, a million copies were sold, reports André Carothers in *Greenpeace*. The album topped the Tass record charts for five months, and was a big hit as well in Australia, the United Kingdom, and West Germany (where it was released under the title *Rainbow Warrior*). All proceeds from sales world-wide support Greenpeace projects in the Soviet Union.

Rock songs and videos certainly reach a different audience than the one first touched by books such as Rachel Carson's *Silent Spring*, which led to strict regulations or bans on some dozen commonly used pesticides after it was published in 1962. They reach the consumers and the parents of tomorrow—those whose actions and attitudes in the next few years can make the difference in progress towards a sustainable society.

* * *

One person who has long recognized the importance of the media's ripple effect is Jon Tinker. For fifteen years, first as the Director of Earthscan and then as founder and President of the Panos Institute since September 1986, Tinker has worked to help journalists ask the right questions about the complex issues surrounding environment, development, natural resources, and population growth. 'Since the publication of the Brundtland Report', he believes, 'the need to make development greener, so that it may become more truly sustainable, has become more widely understood almost month by month.'

In addition to producing books, magazines, syndicated newspaper features, radio programmes, and photographs, Panos has established a regional partnership programme to strengthen the information skills of media and private groups in the Third World. It has supported Third World writers trying to help readers take their first steps towards environmental reconstruction by commissioning case studies, short articles, and sixty in-depth investigative pieces over three years.

As noted in *Our Common Future*, the success of sustainable-development efforts depends on the participation of those affected by the changes to be introduced. The same is true for information. The Panos Institute bases its efforts on the belief that 'if environmental information *for* the South is to have real and lasting impacts, it must increasingly be planned and produced *from within* the South'. In support of this, Panos has helped non-governmental groups launch *Environesia* in Indonesia, the *Environment and Development Journal* in Kenya, *Vikalp* in India (by the Energy and Environment Group), and *Mihikatha* in Sri Lanka (by the Environment Congress).

In a similar effort, the Washington-based Center for Foreign Journalists has been holding seminars with Latin American newspapers for their general-coverage journalists. In late 1988 fourteen reporters from Colombia, Ecuador, and Peru at a Quito workshop issued a challenge to their colleagues throughout the continent: 'We therefore resolve to call on the Latin American press to join in educational campaigns in their communities on environmental topics, to guide governments on this, and promise ever greater space to ecological subjects.'

One of the signatories of the Quito Declaration was Barbara D'Achille, a crusading environmental reporter from *El Comercio* of Lima. 'It's not enough to give the news', she told the workshop. 'You have to explain it.' In March 1989 her weekly page on the environment did just that: it explained the likely impacts of a law on developing the Amazon region that had passed without debate on the final night of the legislative session. Debate on the law reopened. Two months later, D'Achille was killed by Shining Path guerrillas as she was on her way to look at a UN development project in the mountains. The 'Ecologia' page continues. But D'Achille's voice is sorely missed.

Seminars and workshops are also useful for reporters already covering the environment. One such meeting in Bangkok in January 1988, held by the UN Economic and Social Commission for Asia and the Pacific, helped launch the Asian Forum of Environmental Journalists (AFEJ), which consists of eleven national associations. Commenting on this innovative group—the only such regional association in the world, it claims—the UN office notes: 'AFEJ has been a pathbreaking network for the Asian and Pacific region. At a time when many countries are . . . awakening . . . to the realization that they can ignore the destruction of their natural resource base only at their peril, that their fates are inextricably bound with their immediate neighbours, an informed media is the key to regional co-operation.'

One of the first products of this new group was *Reporting on the Environment: A Handbook for Journalists.* In 160 pages this handy guide covers all the basics that any environmental journalist needs, from dealing with technical experts to translating complex information for the lay reader. It includes useful brief guides to such environmental issues as over-fishing, desertification, and acid rain. The *Handbook* has been widely distributed in the region, and translations are already or will soon be available in Bengali, Chinese, and Thai.

AFEJ has a regular news-letter, a clipping service of environmental articles from the region, and an award each year of $4,000 for the best story, series, or radio or television programme on the environment. Recognizing that many people in the region are illiterate, the organization is also planning to develop cartoons, comic books, posters, and other visual media to deliver messages about sustainable development.

As more and more editors understand that 'the environment beat' is not going to disappear, this new breed of reporter will be found on an increasing number of staffs. They might also find it useful to start Forums of Environmental Journalists—covering Latin America, Europe, Africa, and North America.

* * *

Beyond these high-visibility, mass-media efforts to deliver the message of sustainable development, many private groups and even individuals are finding innovative ways to raise public awareness. The environment is turning up not only in animated cartoons and rock videos, but even in mail-order catalogues.

One US outfit, Seventh Generation, sells only 'products for a healthy planet' (items such as cloth diapers, toilet paper from

100-per-cent recycled stock, and Ecover's biodegradable household-cleaning products). And their catalogues do what is unheard of in this business: use up valuable space to include general information about environmental issues, and devote a full two pages to promotions of various groups that customers might consider supporting.

New versions of similar messages—about what each person can do to 'live lightly' on the earth—appear virtually daily. One early leader in the United States was 50 Simple Things You Can Do To Save the Earth by the Earthworks Group in California, whose small office receives mail-bags full of orders by the hour. An astounding 700,000 copies were sold in only four months. Another California group, the Friends of the United Nations, published a useful twelve-page Personal Action Guide for The Earth that has been adapted by groups around the world.

A similar pamphlet, though just on the greenhouse effect, by the World Wide Fund for Nature-UK (WWF-UK) was sent to eighteen million British households. The response was nearly overwhelming: 117,000 people wrote in for more information, and WWF-UK gained some 25,000 new members. One of the purposes of the mailing, the group points out, was to show that the government could—and should—be getting this information out to consumers.

A leading cancer researcher in Sweden, Karl-Hendrik Robert, has taken a similar approach. He started a programme called 'The Natural Step: Toward a Consensus on the Environment' that by the spring of 1989 resulted in every single household in the country—all 4,128,000 of them—receiving a colourful booklet and an audio-cassette. The materials were developed by fifty natural scientists brought together by Dr Robert, and financed by a group that included a bank, an insurance company, a church, a cancer-research foundation, and two trade unions.

The initial effort was so successful at opening the debate on 'what can we all agree are the most urgent environmental problems confronting us, and what can we all agree to do about them?' that a television programme soon followed. The original partners have formed a foundation, and are involved in eleven follow-up projects in Sweden. Twenty regional action committees have been formed to identify pressing local problems and devise solutions; composed of a cross-section of society, they are close in orientation to the provincial Round Tables in Canada (see Chapter 7).

Messages also get delivered by books such as this one, of course, and 1990 has seen a flood of them. In part this reflects their marketability due, in the United States at least, to the twentieth-anniversary celebration of Earth Day. Some twenty books on the environment were published there by late April, including such titles as *Save Our Planet: 750 Everyday Ways You Can Help Clean Up the Earth*, *How to Make the World a Better Place: A Guide to Doing Good*, and *Hints for a Healthy Planet* by Heloise (a syndicated columnist better known for advice on more mundane topics, like how to get wine stains out of carpets).

* * *

The specific messages of *Our Common Future* have appeared in various formats and been recast for varying audiences. Soon after the official launch in April 1987, a *Reader's Guide* was published by Earthscan for the International Institute for Environment and Development (IIED) in London. The booklet was very well received in the United Kingdom, and continues to be in demand, but has not been used in the United States as widely. Perhaps British readers are more accustomed to the hard-hitting photographs used by charities there in fund-raising appeals. The varied reception to this *Reader's Guide* is a good reminder of the importance of locally developed messages.

Not long after the IIED booklet appeared, a 'Tribute to Our Common Future: A Communicator's Guide to the Brundtland Report' was published by the World Media Institute of Ottawa. Along with summaries of each chapter of *Our Common Future*, this sixty-four-page newspaper carried reviews of other recent books, reprints of a few of the comments from public hearings that appeared in the report, and articles on the Canadian response by that point.

Indeed, the World Commission's report generated an enormous amount of interest in Canada for a number of reasons, undoubtedly including the government's financial contribution to the Commission, the involvement on it of two prominent Canadians, and the extensive public hearings held all across the country during a Commission visit in the summer of 1986.

In the United States, interest was slower to build. In early 1989 a *Common Future Action Plan* was prepared for all Sierra Club chapters to help them use the full report, which was called 'an exceptional educational tool'. Readers are led to specific pages in the full report, and suggestions are made about tying recommendations in to local club activities. As the Sierra Club has a half-million members, this booklet can deliver the message of sustainable development to a sizeable and committed audience.

A second US publication about the Commission is *Sustainable Development: A Guide to Our Common Future*, by the Global Tomorrow Coalition. This is a more straightforward summary of the report, aimed at helping Americans change the US position on these issues from one of being 'scarcely more than bystanders', as the coalition put it. The booklet also described how to become involved in the public hearings on *Our Common Future* held in Los Angeles in late 1989 (see Chapter 4).

A final notable delivery of the specific messages of the Commission capitalized on the growing interest of the rock industry described earlier, and on the power of television. On 3 June 1989 a five-hour programme about our common future was broadcast from New York City's Lincoln Center. Between live pickup by satellite and delayed pickup by tape, 'Our Common Future: A Global Broadcast' has so far been seen in fifty-nine countries, with China expected to air the full programme soon.

In a unique combination of information and entertainment, viewers watched singers John Denver in New York and Alexander Gradsky in Moscow compare notes about their concerns for the environment; learned about some sustainable-development success stories around the world; heard performers such as Diana Ross, Elton John, Sissel of Norway, Stevie Wonder, Sting, Joni Mitchell, and Johnny Clegg & Savuko; listened to messages from ten heads of government, including Presidents Bush and Gorbachev, and the UN Secretary-General; heard actor Christopher Reeve talk about global warming, and Sigourney Weaver on her concern about endangered elephants in Africa; and saw ten short interviews with Prime Minister Brundtland about where the world stood now on these issues—and where it should be heading.

*   *   *

Although the Brundtland Report did not have a direct effect on the media, its indirect impact stems from the fact that it pulled sustainable development into the international arena, and on to the agendas of numerous meetings the media has to cover, as indicated throughout this book. The coverage has had to become more sophisticated as it became clearer that there are no simple solutions to these global and local problems of development. It is likely to improve even more in the next decade.

The next step in sophistication will be to stop categorizing all the stories as being about 'the environment', which this chapter itself does in reflection of the media's current interest. But as *Our Common Future* pointed out, 'a world in which poverty and inequity are endemic will always be prone to ecological and other crises'.

Lloyd Timberlake of IIED, in reflecting on this recent revolution in interest at an International Press Institute conference on 'The Environment, the Media, and the Public Interest', put this danger well:

> If we talk and write about environmental problems, we mislead ourselves, and we mislead our readers and listeners into thinking that there is some sort of environmental solution, some technical, scientific answer which can be put in place without involving us. If we can make clear that we are talking about economic problems, political problems, then it is quite obvious that everyone must be involved in the debate and in the solution.

Later, in an interview for this book, Mr Timberlake added:

> We need to get the environment out of the ghetto. Reporters should think to cover environmental causes and/or implications of whatever they are covering. Disaster reporting is a good example of where this has already happened. Disasters used to be viewed as acts of god or nature, but since 1984 journalists have looked further. Most stories on the flooding in Bangladesh in 1988, for example, included some discussion of deforestation in the Himalayas that contributed to the severity of the situation.

Defining sustainable development remains a problem for many who attempt to expose new audiences to the concept. It is still often defined in the negative. As Jon Tinker points out in the *New African*:

> By bitter experience many Africans have come to understand all too well what is *not* sustainable. . . . The woman in northern Ghana who has seen the water level in the village drop year by year; the boy herding goats in Kordofan who has to go for miles to find any pasture at all; the young girl walking an extra three miles in Zimbabwe because the nearest waterhole has dried up; the fisherman on the Nigerian coast whose livelihood has vanished beneath coastal oil pollution—far too many Africans have learned the hard way what is *not* sustainable.

The problems caused by such development receive more coverage now than before. But we also need to read about, and to see, success stories. Stories like those in *Against All Odds* and *Banking the Unbankable*. These Panos collections of articles by Third World journalists document how community groups in Asia and Africa are solving environmental problems and how community credit schemes, mostly created by women's groups, are working in eleven countries. Commenting on the role of such articles, Donatus de Silva notes: 'Very often the individual feels powerless in the face of huge bureaucracies. Awareness about what ordinary people could do to change their circumstances could turn despair into a commitment to a better future.'

*   *   *

The larger issue of the media's role in working towards our common future concerns the acceptability of a journalist 'taking a stand' on issues. To return to *Time*'s 'Planet of the Year' cover story, the magazine listed eight steps the US government should take: raise the gasoline tax, toughen auto fuel-efficiency requirements, encourage waste recycling, promote natural-gas usage, encourage debt-for-nature swaps, support family planning, ratify the Law of the Sea treaty, and make the environment a summit issue. Six months later, their last idea had already come to pass.

Recommendations like these are supposed to stay on the opinion pages, and the ripples from *Time*'s bold stance are still being felt. The propriety of such a list was a topic of lively debate at the September 1989 Smithsonian gathering. Charles Alexander of *Time* was not ashamed to say that they had crossed the boundary from news reporting to advocacy at the magazine, as of 2 January 1989. *Washington Post* editor Ben Bradlee wanted to be a bit more coy: 'I don't think there's any real danger in *doing* what you suggest, but there's some danger in *saying* you are doing it because as soon as you say "I'm no longer interested in news, I'm interested in causes" you've got a whole kooky constituency to respond to.'

Frances Cairncross of *The Economist*, on the other hand, disagrees: 'In my experience, once newspapers start preaching, readers stop reading.' Indeed, the UK workshop where she made this point was on 'Awareness Raising and Public Participation'. Those attending concluded that 'the media has no direct role in promoting sustainable development other than reporting information'.

The argument on this issue has often been cast in terms of the ethics of 'advocacy journalism'. But at the Smithsonian gathering, Barbara Pyle of Turner Broadcasting argued that what is being advocated is survival:

> I believe that we have a moral obligation to inform the public of any life-threatening situation on this planet. If someone was about to be pushed off of a bridge, would you run up and tell them that they were to be pushed off a bridge or would you watch them be pushed off?
>
> The planet is about to be pushed off the bridge, in my opinion. . . . Either we're part of the solution as a group or we're part of the problem. Who will be held accountable in ten, twenty, thirty, or forty years if we do not have an informed constituency to make these difficult decisions and the decisions are not made?

Captain Planet cartoons, regular 'Ecologia' pages in newspapers and 'Environment' columns in magazines, *TV Viva* in the streets of shanty-towns and *Greenpeace* in 2.4 million readers' hands, feature movies on environmental martyrs, expanding associations of environmental journalists. All these means of delivering the message are small steps towards a better-informed constituency. And maybe signs that some people in the media are starting to ask how they can be part of the solution.

7

New Ways of Thinking

When the Canadian Minister of Environment, Lucien Bouchard, addressed the UN General Assembly in October 1989, he talked about the impact of the Brundtland Report:

> In articulating the concept of sustainable development, the Commission changed forever the way we think about the environment. . . . We must now take the next step and translate sustainable development into reality.
>
> We should do this without illusion. Let there be no mistake. Sustainable development is a radical concept, not the *status quo* in a new package. We are talking about fundamental changes in the developing world and, even more importantly, in industrialized countries. Sustainable development need not entail large capital outlays. But, as pointed out in the Brundtland Report and confirmed in the UNEP definition of sustainable development, it requires a new way of thinking about future development.

Echoes of Edward DeBono, a researcher in thinking patterns and skills whose books on 'lateral thinking' were popular in the late sixties. DeBono explained his approach as 'breaking out of the old, self-perpetuating patterns and generating new ways of looking at things'. Especially popular were games to train lateral thinkers. You would be given the last line of a story—'if he had seen the sawdust, he wouldn't have died', for example—and challenged to figure out the rest of the tale. You could quiz those who knew the answer as much as you wanted, but only with questions that could be answered yes or no. Eventually, usually after hours of working hard to think laterally, something would click, and the complex circumstances that led to that last line would suddenly become clear.

If sustainable development is the last line, what questions should we ask to get us there? This chapter looks at some lateral thinking going on in research institutes, governments, and citizens' groups— new ways of thinking about appropriate technology, realistic accounting of natural-resource use, financing mechanisms, institutional change, and how to motivate people to move more quickly towards the answer: sustainable development.

*   *   *

In *Our Common Future*, the World Commission on Environment and Development noted that 'not enough is being done to adapt recent innovations in materials technology, energy conservation, information technology, and biotechnology to the needs of developing countries. These gaps must be covered by enhancing research, design, development, and extension capabilities in the Third World.'

Since 1983 an organization in India called Development Alternatives has been doing some lateral thinking on that very issue. Based in New Delhi, the group carries out innovation, production, and marketing activities throughout the country, in arrangements with local franchised entrepreneurs. Instead of asking what was wrong with people who would not use the many inventions that have flowed from laboratories for years, researchers asked what was wrong with the technologies.

The mission of Development Alternatives is 'to promote environmentally sound development and to achieve the mass deployment of appropriate technologies'. To do this, it relies on the independent sector, which it calls a new type of organization. The independent sector must combine 'the inventiveness usually found in a university, the social objectives of a voluntary organization, the motivation of the private sector, the reach of the government— and the participation of the people'.

Ashok Khosla, President of Development Alternatives, was the first Director of India's Office of Environment. He notes that perhaps 15 million people in that country benefit from modern technology, but that another 650 million have neither purchasing power nor physical access to more than a very few of the products. Appropriate technology must meet the needs of those millions. It aims to improve the quality of their lives, draws on existing resources and skills in an area, and does not destroy the resource base on which it depends. By definition, then, an alternative product cannot be designed in the abstract in a research laboratory, Mr Khosla writes; it springs from indigenous creativity, in response to local needs and possibilities.

People have been saying that was needed, however, at least since the first development-assistance grants were awarded. In India, Development Alternatives has combined the theory with practice by working with local people to take the best of technological innovations and adapt them to local conditions and needs. The franchise networks it has set up aim to 'combine the power, financial resources, and economies of scale of the big corporation and the responsiveness of the small, local unit'.

Examples of the products investigated and eventually marketed by the group include bio-gas and portable metal cooking-stoves, water pumps, low-cost and efficient hand-looms, food-storage bins, paper- and board-making equipment, and a soil-block press. In fact, their Delhi headquarters was constructed out of bricks from local soil, serving as a showcase for the application of appropriate technology. Inside, researchers sit at computer terminals, tracking technological developments they might want to adapt to local needs and conditions in India.

Development Alternatives expects in 1990 to have a staff of more than 200 scientists and specialists operating from a dozen centres throughout India. An affiliated organization, Technology and Action for Rural Advancement, will be manufacturing and marketing some fifteen different products, which Development Alternatives claims makes it the largest appropriate-technology organization in a developing country. Its investments in the research and development of appropriate technologies by 1990 will have passed 20 million rupees ($1.2 million).

Through some lateral thinking, these researchers, entrepreneurs, and development activists in India are working to answer the question: How can we adapt the best technological developments in the world to meet the needs of a vast and growing market—the rural and urban poor in developing countries?

*   *   *

How would society measure the value of bricks from a shop set up by Development Alternatives versus that of bricks transported from a distant factory built on land where a forest recently stood? In traditional accounting, the money spent to cut down the forest, to build the factory, and to ship the bricks would add value to the national income. But the loss of the timber and the forest's other values remains uncounted. And the benefits provided by the Development Alternatives shop—sustaining employment locally, using local resources, and increasing local self-sufficiency—have no quantifiable value.

Over the last few decades, some economists have been working on a new way of thinking about this key issue. How can accounts of a nation's income include measures of the degradation of natural resources they are based on? Economists attach value, for example, to the activity of cleaning up pollution and to the health-care industry that deals with the ill effects of dirty air. Yet nowhere in the gross national product can they indicate that it is harder from one year to the next to breathe deeply—and safely. You do not have to understand the intricacies of transaction flows, tangible capital, stocks of assets, or liabilities to see that this makes little sense (see Box 7–1).

The Brundtland Commission considered this problem in its discussion of the need to change the quality of growth. The Commissioners concluded that 'the process of economic development must be more soundly based upon the realities of the stock of capital that sustains it. . . . In all countries, rich or poor, economic development must take full account in its measurements of growth of the improvement or deterioration in the stock of natural resources.'

This adjustment seems so obvious and logical, it is hard to understand how it could have been ignored for so long by statisticians. National-income accounts, it turns out, were devised during the Great Depression, when natural resources seemed abundant and economists had more pressing problems on their minds.

In testimony before a US congressional committee in 1989, Robert Repetto of World Resources Institute (WRI) explained the bottom-line importance of undertaking such revisions: 'If depletion of natural resources can no longer masquerade as income growth, governments tempted to engage in environmental deficit financing will be less able to hide behind a reassuring screen of economic indicators. Policies that promote destructive and wasteful uses of natural resources will no longer be justified so easily as necessary for economic growth.'

Box 7–1. Today's Poor Accounting Practices

Natural resource assets are legitimately drawn upon to finance economic growth, especially in resource-dependent countries. The revenues derived from resource extraction finance investments in industrial capacity, infrastructure, and education. A reasonable accounting representation of the process, however, would recognize that one kind of asset has been exchanged for another, which is expected to yield a higher return.

Should a farmer cut and sell the timber in his woods to raise money for a new barn, his private accounts would reflect the acquisition of a new asset, the barn, and the loss of an old asset, the timber. He thinks himself better off because the barn is worth more to him than the timber.

In the national accounts, however, income and investment would rise as the barn is built, but income would also rise as the wood is cut. The value of the timber, less that of any intermediate purchases (e.g., gas and oil for the chainsaw) would be credited to value added in the logging industry. Nowhere is the loss of a valuable asset reflected.

This can lead to serious miscalculation of the development potential of resource-dependent economies by confusing gross and net capital formation. Even worse, should the proceeds of resource depletion be used to finance current consumption, then the economic path is ultimately unsustainable, whatever the national accounts say.

If the same farmer used the proceeds from his timber sale to finance a winter vacation, he would be poorer on his return and no longer able to afford the barn, but national income would only register a gain, not a loss in wealth.

Source: Robert Repetto *et al.*, *Wasting Assets: Natural Resources in the National Income Accounts* (Washington, DC: World Resources Institute, 1989).

The call for a new economics has been heard for some time in environmental circles, in the writings of, for example, Paul Ekins, Hazel Henderson, and James Robertson. For about the last fifteen years, a few more-traditional economists have been working on revising accounts in Canada, France, Japan, Norway, and the United States. By the mid-eighties, the plea to rethink the way national income is measured was heard by a broader audience. And in 1989 WRI published *Wasting Assets: Natural Resources in the National Income Accounts*, by Dr Repetto and several other development economists. In addition to discussing the basis for changing the way governments measure income, they took a hard look at the accounts in Indonesia.

The authors estimated the depreciation of petroleum, timber, and soil resources for 1971–84. Gross domestic product in Indonesia appeared to increase 7.1 per cent a year during this period, but by their calculations the 'net' domestic product rose only 4.0 per cent annually. If an estimate of the depletion of the country's other valuable natural resources—such as coal, copper, tin, nickel, natural gas, and fisheries—had been included, the difference between appearance and reality would have been even greater. Their conclusion: 'Conventionally measured gross domestic product substantially overstates net income and its growth after accounting for consumption of natural resource capital.'

In August 1989 this view was endorsed in the United Kingdom in *Blueprint for a Green Economy*, a study by David Pearce, Anil Markandya, and Edward Barbier of the London Environmental Economics Centre, a new research group formed by the International Institute for Environment and Development and University College. The report had been commissioned by the UK Department of the Environment, which called it a 'benchmark study' in its own report a month later on progress in implementing sustainable development. The government noted that 'mechanisms are urgently required which will enable sustainable development to be measured and progress towards its achievement assessed. The UK regards this challenge as a major priority.'

The Pearce Report, as it quickly became known, reviewed the meaning of sustainable development in economic terms and the evolving efforts to adjust national-income accounts. Two different approaches have been followed. Canada, France, and Norway make some attempt to put resources into physical terms: tonnes of fish caught and added to reserves, hectares of land developed, tonnes of sulphur dioxide spewed into the air, and so on. These are generally presented in satellite income accounts, not incorporated in the main measure of national wealth. The World Bank, the UN Environment Programme, and some bilateral aid agencies are considering how to help developing countries establish such accounts; research has started on estimates for Costa Rica, the Philippines, and China.

The second approach, akin to the WRI analysis of Indonesia's accounts, tries to put a monetary value on natural-resource restoration and degradation, and incorporate that in the main figure cited as a nation's income growth. In 1989 West Germany announced it was starting to work on such a comprehensive income figure. Dr Pearce notes that the Japanese government already tries to evaluate the loss of human welfare caused by environmental damage, and then uses that to come up with a net national-welfare number.

The bottom line of this comprehensive *Blueprint for a Green Economy* is a recommendation to the UK government to develop satellite accounts in either physical or monetary terms, or both, as a way to work towards a measure of sustainable income, 'the flow of goods and services that the economy could generate without reducing its productive capacity—i.e. the income that it could produce indefinitely'.

Pressure for revised accounting systems has been taken to the global level through appeals to the UN Statistical Commission. This body is currently considering changes in the system of national accounts that the United Nations keeps for all countries, a revision it undertakes only once every twenty years. According to Robert Repetto, a preliminary decision has been taken to make no fundamental changes in the existing accounts, a decision he suggests the Commission should reconsider. The lack of a UN imprimatur for integrated accounts makes the development of comprehensive satellite accounts in as many nations as possible all the more imperative.

National-income accounts may sound like a pretty obscure subject, of interest to a few academics perhaps and to accountants, a profession that is often the butt of comedians' jokes. But the wrong signals that policy-makers have been getting for fifty years are no joke. Instead of just accepting the 'indicators of progress' issued annually by governments and the United Nations, these lateral thinkers are saying: Step back, look at what those numbers are ignoring. By giving us a more realistic idea of where we are, work like this can help us figure out where we—and the planet— are going.

*   *   *

It does not take much lateral thinking to come up with the toughest question of all, although it may take quite a bit to get the answer. The question raised most often in international negotiations, national debates, and public meetings is: How are we going to pay for all this?

Where do we get the funds to stop soil erosion, educate those here today as well as the next generation, reforest the earth, halt desertification, help create income-generating jobs in the Third World, clean up the air and water, develop renewable energy and CFC substitutes, and provide family planning to all who want it? In other words, how do we finance sustainable development?

In 1987 the Brundtland Commission noted that 'given the current constraints on major sources and modes of funding, it is necessary to consider new approaches as well as new sources of revenue for financing international action in support of sustainable development'. In late 1989 the Executive Director of the UN Environment Programme not only concurred, he considered the problem to be the major challenge we face. In a letter with comments for this book about priorities in the nineties, Mostafa Tolba wrote: 'There is no doubt financing solutions to the global environmental problems will be *the* major issue of this decade. It will be the centre-piece of the United Nations 1992 Conference on Environment and Development. New financing will have to include innovative ideas about revenue generation and financial building.'

Some environmental groups started to think laterally about this problem a few years ago. They looked at one major problem facing numerous Third World governments—external debt—and at the simultaneous need to protect threatened species and ecosystems while enhancing local development opportunities, and said why not tackle both problems at once.

The results have been called debt-for-nature swaps. The first was arranged in Bolivia. A US-based environmental group, Conservation International, bought $650,000 worth of the government's debt from a private bank (discounted to $100,000) and cancelled it; in return, the government agreed to give the highest legal protection it could to the nation's Beni Biosphere Reserve. It also set up an endowment worth $250,000 that would go towards managing and protecting the area. This reserve and its surrounding buffer zone cover 2.7 million acres (about 1 million hectares), and had been under increasing pressure from haphazard development in the region. In effect, the Bolivian government is still paying off its debt, but now the interest is smaller, can be paid in local currency, and goes into a local fund that supports development of a sustainable forest-use plan.

Since the initial swap in 1987, Conservation International

and other private environmental groups have worked with the governments of Costa Rica, Ecuador, Madagascar, and the Philippines on variations of this same theme. The groups raising the funds in industrial countries generally work with established local conservation organizations. In Ecuador, for instance, Fundacion Natura received in late 1987 government bonds in local currency worth $1 million after World Wildlife Fund-US bought debt worth that much on the open market for $354,500. The interest from these local bonds will support protection-and-management activities in six local parks, and when the bonds mature at the end of the decade, they will provide a nice sum—in local currency—for a Fundacion Natura endowment fund. Subsequent swaps wrote off a further $9 million of Ecuadoran debt.

To date, debt-for-nature swaps have erased about $100 million of external debt. If the World Bank and bilateral aid agencies become involved, as some are urging they should, these exchanges could have a much broader impact. At the moment, many commentators point out that $100 million is a mere drop in the bucket of Third World debt, a bucket it would take almost $1,200 billion to fill. Nevertheless, the swaps are important for the local areas that are better protected as a result, for the environmental groups in the developing country that become involved in overseeing the programmes, and for the example of lateral thinking they provide. When faced with two seemingly insoluble problems, try tackling both at once.

Another type of finance being much discussed taps a source that is anything but new: taxes. In late 1989 the US Congress passed a law to tax CFCs and halons, the villains in the eroding ozone-layer problem. Producers, users, and importers will face a tax based on the particular chemical's ozone-depletion potential, a number specified by the Montreal Protocol (see Chapter 2). The $4.5 billion the tax is expected to generate over five years will go towards reducing the US budget deficit—a worthy cause, no doubt, but not of much direct help to the ozone layer.

In fact, this is another drop in another very large bucket. Vast sums are needed in an international pot that developing countries can draw from in order to avoid repeating the mistakes of industrial countries. The Third World simply cannot afford on its own to get off the industrialization path laid by the North. One way to handle this is to set up a climate or atmosphere fund, an idea first brought to international attention at a conference in Toronto in June 1988. This could be financed through 'carbon taxes' in each nation.

Such a tax could be levied on fossil fuels in relation to the carbon dioxide contributed to the atmosphere when they are burned. Using coal would cost the most, oil would be next, and natural gas the least. According to the World Resources Institute, a tax pegged at the level of 10 per cent on coal could generate more than $25 billion a year in revenues if introduced throughout the countries belonging to the Organisation for Economic Co-operation and Development. Combining this with a charge for CFC use could create an overall 'greenhouse gas tax', which some commentators have called for.

The Pearce Report in the United Kingdom noted that such a tax would encourage individual householders and businesses to switch to less carbon-intensive fuels and to conserve energy. Whenever it is introduced, it should be combined with exemptions or provisions that would avoid penalizing those least able to afford conservation efforts: the poor and the aged. In considering the potential impact of a carbon tax, David Pearce concluded that it merits consideration for any global convention on carbon emissions, but that it is likely to have negligible impact on global warming if introduced in only one country.

Nevertheless, some governments are considering just such legis-lation—and one has already passed it. In the Netherlands, fossil-fuel use was taxed according to its carbon content as of January 1990. The revenue is expected to reach 150 million guilders ($75 million) a year. Similar proposals are being debated in various offices and legislative chambers of the United States and the United Kingdom. And Sweden expects a proposed tax to be enacted soon, although an unexpected change in government in February 1990 may slow the process down. If approved, that tax is expected to result in annual carbon-dioxide emissions that are five to ten million tonnes lower by the year 2000 than they would be otherwise.

Another way to fill the coffers of a climate fund would be through direct contributions not tied to fossil-fuel use, deforestation rates, or any other specific climate issue. In its April 1989 report on follow-up to the World Commission, the government of Norway indicated that if other industrial countries supported the idea, it was prepared to allocate 0.1 per cent of its gross national income to an international fund for the atmosphere associated with the

United Nations. In 1989 this would have meant a donation of 700 million kroner ($107 million). Whether such a fund will be set up, and how it would be managed, is the subject of considerable debate in the world's capitals.

In September 1989 Rajiv Gandhi, then Prime Minister of India, took this idea of a fixed contribution one step beyond just concern about the climate. At the Belgrade summit meeting of the non-aligned nations, he proposed that a planet protection fund be established (see Box 7-2). Using the same 0.1 per cent of national income figure as the Norwegians, Mr Gandhi estimated that a fund of some $18 billion could be established.

The Prime Minister's emphasis on the need for 'conservation-compatible' technologies in developing countries is echoed in some points that Dr Tolba made in his letter: 'If developing countries are expected to follow, for example, the discipline of the Montreal Protocol and use more costly CFC-substitutes, then they must have fair access to affordable environmental technologies. Legally binding global agreements on technology thresholds and enforcement mean little unless they are re-inforced with legally binding agreements on North-South financial and technical co-operation.'

Within weeks, the idea of a planet protection fund was endorsed by the Commonwealth Heads of Government at a meeting in Malaysia. It was also supported by Soviet Foreign Minister Eduard Shevardnadze, writing in *Literaturnaya Gazeta*. He added: 'We have grown used to the fact that capital is moving mostly out of the developing to the developed world, and not vice versa. We must revert this trend so that at least a part of these means return to the developing countries, say, as ecological aid.' Financing, Mr Shevardnadze noted, must be one of the key issues at the 1992 Conference on Environment and Development in Brazil.

The proposals to date depend almost entirely on the good will and enlightened self-interest of the wealthier, industrial countries. What if this is not enough? What if the world has waited too long to ask for contributions to a planet protection fund? In the very final pages of *Our Common Future* the Commission dealt with this difficult and delicate issue—should all countries but the very poorest be forced to contribute to our common future? The Commissioners' recommendations are worth repeating here:

Box 7–2.   Rajiv Gandhi's Proposal for a Planet Protection Fund

We recognise that sustainable development begins at home, that the costs of development must integrally include the costs of conservation, which, if not paid for now, will be exacted from the development process later or elsewhere. The environment is not, and should not be made, yet another issue of North–South contention. Nevertheless, in any global endeavour, the legitimate concerns and interests of the developing countries must be fully met. We cannot isolate environmental protection from the general issues of development. We need positive and supportive measures to assist the developing countries in exploring and implementing environmentally benign policies of development. Environmental degradation is an issue which transcends not only national boundaries, but also, in some ways, narrowly conceived national interest. We are ready to do our part. . . .

We need a global effort to bring within the easy reach of all, developing and developed, the technologies that exist and are yet to be developed to combat pollution and environmental degradation. We cannot leave these matters to the mere ploy of market forces. Those with inadequate capacity to pay for environmentally sound technologies would then be left with no alternative but to let development proceed without due regard for the environment. Equally, those who are denied access to such technologies would have no option but to fall into the same trap. It is not only these countries that will pay the price of environmental neglect, it is a price that will probably have to be paid by the world as a whole and by future generations. The international community has a common stake in sustainable development. We need a global effort to ensure access to environment-friendly technologies and the funding of research and development into such technologies.

The search for other, and especially more automatic, sources and means for financing international action goes almost as far back as the UN itself. [In] 1977 . . . when the Plan of Action to Combat Desertification was approved by the UN General Assembly . . . governments officially accepted, but never implemented, the principle of automatic transfers. . . .

Since then, a series of studies and reports have identified and examined a growing list of new sources of potential revenue, including: revenue from the use of international commons . . . taxes on international trade . . . and international financial measures.

Given the compelling nature, pace, and scope of the different transitions affecting our economic and ecological systems as described in this report, we consider that at least some of those proposals for

With these ends in view, I propose the establishment of a Planet Protection Fund (PPF) under the aegis of the United Nations. The Fund will be used to protect the environment by developing or purchasing conservation-compatible technologies in critical areas which can then be brought into the public domain for the benefit of both developing and developed countries. All technologies over which the Fund acquires rights will be made available gratis, and without restriction, to all constituent members of the Fund. I would wish to stress that contributors to, and beneficiaries of, the Fund would include not only developing countries but also the industrialised countries. We would wish to work towards universal membership of the Fund.

We propose that all constituent members of the Fund, developed and developing, contribute a fixed percentage of their Gross Domestic Product (GDP) to the Fund, with exemption but full access granted to the least developed countries. The annual contribution to the corpus of the Fund would be around $18 billion at as low an average contribution as of 0.1% of GDP. That is, for environment-related work, the international community would have at its disposal as significant a sum as $18 billion a year, if only each country were willing to part with but a one-thousandth part of its GDP.

Such a Fund would become the fulcrum for a truly cooperative global endeavour to measure up to a problem of global dimensions and global implications. Such a Fund would be proof of our commitment to saving all creation and our planet Earth.

Source: Excerpted from Address by Rajiv Gandhi, Prime Minister of India, at the Ninth Conference of Heads of State or Government of Non-Aligned Countries, Belgrade, 5 September 1989.

additional and more automatic sources of revenue are fast becoming less futuristic and more necessary.

*   *   *

On many people's lists of challenges in the nineties, the second item—right behind money—is the need for institutional change. As the Commission noted: 'The integrated and interdependent nature of the new challenges and issues contrasts sharply with the nature of institutions that exist today. These institutions tend to be independent, fragmented, and working to relatively narrow mandates with closed decision processes. . . . The real world of interlocked economic and ecological systems will not change; the policies and institutions concerned must.'

This means more than just the agriculture ministry consulting with the forestry ministry before it sets targets for land to come under the plough, for example. And it means more than just four ministries jointly preparing an environmental-policy plan, as was done in the Netherlands, admirable though that process was. It means all the stake-holders, as Canadians often put it, considering the problems and the solutions together.

Canada, in fact, was one of the first to take some lateral steps on this issue. In September 1987, before *Our Common Future* was even presented to the UN General Assembly, a National Task Force on Environment and Economy published a report on the implications for Canada that has been called a historic statement. Media and public interest in this report was 'overwhelming', according to the Task Force; 50,000 copies were distributed throughout the country.

The statement was historic, in part, because of who signed it— a group that probably included more representatives of the various stake-holders than had ever collaborated on a consensus document before. Included were the Presidents of Alcan Aluminum and the Canadian Petroleum Association, the Executive Vice-President of Inco Limited, the Chairmen of Dow Chemical Canada and of Noranda Forest Inc., a representative of the Ecology Action Centre, and the Minister of Environment for Canada and five provincial environment ministers.

The group had been set up by the Canadian Council of Resource and Environment Ministers as a direct result of the Brundtland Commission's visit to Canada the preceding year. 'Our main objective', the Task Force noted, 'is to promote environmentally sound economic growth and development, not to promote either economic growth or environmental protection in isolation.'

Of its forty recommendations, the group identified as one of the most important the suggestion for 'a new cooperative initiative to integrate economic and environmental planning through the participation and debate of senior decision makers in every province and territory and at the national level in Canada'. These 'Round Tables' should be chaired by individuals appointed by the First Ministers of each jurisdiction, the Task Force suggested, and it concluded that their implementation and success were fundamental to the achievement of environmentally sound economic development in Canada.

Thus was the National Round Table on the Environment and the Economy (NRTEE) born, in answer to the Task Force's challenge that 'in a new era . . . a full partnership of governments, industry, non-governmental organizations and the general public must guide us through an integrated approach to environment and economy'. Prime Minister Brian Mulroney appointed to the NRTEE twenty-five Canadians from industry, academia, research institutes, environmental groups, and the federal and provincial government—including the Ministers of Environment, of Finance, and of Industry, Science and Technology.

The NRTEE set up a mixed bag of strategic and specific committees: decision-making, socioeconomic incentives, foreign policy, recycling, and education and communications. Following its inaugural meeting in June 1989, opened by the Prime Minister, NRTEE members planted a tree in a park in Ottawa—a symbol, they claimed, of the early stages of development, whether of nature itself or sustainable development.

As recommended in the 1987 Task Force report, Round Tables have also been set up in the ten provinces of Canada, plus in one territory, drawing on similarly diverse groups of people and aiming to bring to the same table those with traditionally competing interests. Overall, these initiatives in Canada provide one of the few examples of lateral thinking on institutions since *Our Common Future* was published. As this new form of participatory democracy is just getting under way, however, the long-term impact and staying power is difficult to judge.

Somewhat similar initiatives exist in a few other countries, though none yet is as far-reaching. In Denmark, as part of its Plan of Action for Environment and Development, the government is encouraging the establishment of 'green municipalities' that will, among other activities, 'provide examples of how environmental problems can be addressed in new, preventive, and more effective ways'. It is hoped this will help others stop resource and environmental problems before they arise. Areas of possible campaigns include health, life-style choices, fisheries management, transportation, water and energy conservation, tourism, and education.

Nine municipalities have set up councils so far, and many others have shown interest, according to Niels Bo Sørensen of the Ministry of the Environment. With the objective of involving as many citizens as possible in planning projects, the green municipality is seen as a 'dynamo and source of inspiration'.

In the United States, one recently formed group also tries to bring various stake-holders together. The Source Reduction Council of the Coalition of Northeastern Governors was set up in September

1989 to help devise goals for reducing packaging waste, eliminating toxic materials in packaging, and developing consumer-education programmes. Included in the group are representatives of Environmental Action, the Natural Resources Defense Council, National Audubon Society, Conservation Foundation, Scott Paper Company, Sears Roebuck, Mobil Chemical Corporation, Pepsi-Cola, and the Campbell Soup Company. Within two months a committee had drafted model legislation to ban the use of heavy metals in packaging; eight of the nine governors announced right away they would introduce the council's bill.

Perhaps a few more initiatives like this will lead to a US Round Table on the Environment and the Economy. A December 1989 speech by Du Pont's new chairman, E. S. Woolard, hinted at such a development: 'We need a constructive new coalition of environmentalists and industrial leaders, along with legislators and journalists, to work together responsibly on the environmental challenges we face.' He was echoed a month later by Gro Harlem Brundtland, speaking in Moscow: 'We need to build stronger coalitions, to bring all sectors of our societies together in a committed effort to tackle the root causes of our predicament—poverty, over-population, mismanagement, and abuses of the right of our citizens to live in a healthy world environment.'

In Australia, some new thinking about government institutions is worth noting. In *Our Common Future*, the Commission called on governments to consider developing a 'foreign policy for the environment'. Prime Minister Bob Hawke's government did just that—and in response appointed the world's first Ambassador for the Environment, to 'give Australia a strong and clear voice in the important international debates on environmental issues now taking place'. The Prime Minister announced in July 1989 that Sir Ninian Stephen, former Governor-General of Australia, would take up this new post.

The appointment was greeted in some corners as an attempt by the government to woo middle-ground voters who are increasingly concerned about the environment. Whether that was the motive or not, having one person report directly to the Prime Minister on diplomatic developments regarding climate change, the ozone layer, deforestation, Antarctica (an issue of particular concern to Australians), and everything else that will surface in the nineties is bound to help the left hand of government know what the right hand is doing.

In Europe, co-ordination along these lines will be aided by a mid-1989 decision of the twelve members of the Community to establish a European Environment Agency. The new office intends to build on existing networks and institutions to collect information initially on air and water quality, soil erosion, and nature conservation. Periodic 'state of the environment' reports for the region are expected. European governments that are not members of the Community will be invited to join in on some of the programmes, in further acknowledgement that borders mean little to pollution.

Central American governments took a similar step towards greater environmental co-ordination in late 1989. The Presidents of Costa Rica, El Salvador, Guatemala, Honduras, and Nicaragua decided to set up a Commission on Environment and Development. The agreement they signed notes that they 'are convinced that to assure improved quality of life to the Central American people, it is important to support respect for the environment within a framework of a sustainable development model . . . and are sure that the regional rationalization of natural resource use is a fundamental ingredient for the achievement of a lasting peace'.

The new office will act as a sort of regional environmental protection agency; it hopes to work with the members on such issues as setting effluent standards and establishing a tropical forestry action plan for the region. The Presidency will rotate among the five Ministers of Environment, with Costa Rica's Alvaro Umaña taking on responsibility for the first year. As this initiative has just been announced, like the European one, it is hard to know what the impact will be.

These moves towards better national and regional management are heartening. But Commonwealth Secretary-General Shridath Ramphal summed up the larger management problem facing the world at a Royal Society of Arts lecture: 'We are overburdened by several centuries of sovereignty of the nation state and all the adversary systems built round it. How do we get beyond global planning to global management until we shed some of that baggage from those centuries? We have to do it together.'

The United Nations is still the best forum for the co-operation needed to secure our common future. It brings its own baggage, but the load seems to have lightened considerably in the last few years. A flurry of proposals have been made recently on how to tap the unique resources of the United Nations and the office of the Secretary-General. During the 1989 General Assembly, for example, Austria's Minister of Foreign Affairs Alois Mock suggested a force of 'UN Green Helmets' to complement the Blue Helmets that have been so successful in peace-keeping assignments around the world.

These new troops could help prevent environmental disputes from arising, or could help settle existing ones. They could assist in investigations of any environmental situations, using the term broadly, that the Secretary-General considered a threat to international peace and security or to the 'global commons'. In a variation on this theme, Ken Piddington of the World Bank has proposed that UN green-keeping forces consist of disinterested negotiators who could moderate discussions on various global environmental topics.

In January 1990 President Gorbachev put forth another idea for international green-keeping: 'Perhaps it really would be worth setting up a sort of international "green cross" which would come to the aid of states in the event of ecological disaster. The USSR's proposal to set up a center for emergency ecological aid under the United Nations is in this vein.'

More basic change within the United Nations itself is also being discussed. Proposals have been made to set up a new ecological security council, to reinvigorate the little-used Trusteeship Council, to have the full Security Council hold a special session on the environment at regular intervals, to set up a continuing Independent Commission, or to strengthen the mandate for change given to the UN Environment Programme, which could act on behalf of the Secretary-General or of one of these new councils. Several of these ideas appeared in *Our Common Future* as well. The difference is that now they are heard in speech after speech; they are being talked about both formally and informally by national leaders as well as international diplomats.

These various proposals on global management are bound to be widely discussed during the months before the 1992 conference in Brazil. All of them aim at modifying and working within the existing UN organization in order to better manage the world's environmental affairs. One recent statement, however, suggested a step that would take that system into a different realm.

In March 1989 seventeen heads of government and representatives of seven other industrial and developing countries met in The Hague to consider whether the threats to the earth's atmosphere were so grave that new principles of international law and more effective decision-making and enforcement mechanisms were called for. They decided that indeed they were grave, and that indeed a new way of thinking was called for. In the Declaration of The Hague, those attending acknowledged and agreed to promote:

> The principle of developing, within the framework of the United Nations, new institutional authority, either by strengthening existing institutions or by creating a new institution, which, in the context of the preservation of the earth's atmosphere, shall be responsible for combating any further global warming of the atmosphere and shall involve such decision-making procedures as may be effective even if, on occasion, unanimous agreement has not been achieved.

To translate this 'diplomatese', some environmental problems have become so global in scope and impact that a supra-national environmental protection agency may have to be created that could decide to fine or otherwise penalize members who did not take whatever steps the majority deemed necessary. The declaration went on to suggest that the International Court of Justice could be used to settle any disputes about the new authority's decisions. As Mrs Brundtland noted at the end of the meeting: 'The principles we endorsed are in fact radical. But any approach that is less ambitious will not serve us.'

Since March 1989 a total of forty governments have endorsed The Hague Declaration. Although it is not a binding document, it is an important step in a new direction. It means that forty governments have now indicated they would contemplate giving over some of their sovereignty. This is the heaviest piece of baggage the nations of the world need to shed.

*   *   *

Technology appropriate to the task . . . a realistic picture of where we are . . . additional funds . . . new institutions. These are all essential components of the transition to a sustainable society. But also needed is a whole new way of thinking about something that cannot be measured, a problem that cannot just be 'solved': motivation.

'We know we have to change, but the problem is change itself', said former Commissioner Maurice Strong in an interview done for this book about priorities for the nineties. 'We need to overcome the inertia of our behaviour patterns.' The quite human reaction to hearing from a physician that we must stop smoking or cut back on foods with cholesterol provides an interesting parallel, he noted. We know the behaviour is hastening our death, or is quite likely to, yet it is hard to stop.

People are starting to see the prognosis for the planet's health in the same light. But what will it take to change the behaviours that are contributing to that poor prognosis? The whole world need not live an austere life, but changes in life-style will certainly be required. 'We must create a demand for sustainable development', Mr Strong believes.

Earth Day 1990 may have provided some of the needed impetus, as millions gathered world-wide on the twentieth anniversary of an event that sparked the American environmental movement. People planted trees in Mauritius, rode bicycles in a demonstration in India, reached the summit of Mount Everest in an attempt to clean up the remotest spot on earth littered with garbage, and listened to scores of speakers in hundreds of parks around the world. The 1970 event led to the creation of the US Environmental Protection Agency. Twenty years later, the ground-swell from people demanding that politicians pay attention to the health of the earth may supply the needed spark to the global environmental movement, proving that indeed you can change the world.

Motivating politicians to change—to raise taxes, for example, which they all dread to suggest—is difficult. Harder still, however, may be motivating people in industrial countries to change their consumption patterns, to live more lightly on the earth, as some put it. It may need a new way of thinking about the earth and our responsibility to it. Whether the appeal is to Gaia, to a 'higher authority', to God, or to the spirits of the earth, people are starting to see that the world will not change its ways just through appeals to the cold, hard facts of environmental decline.

'We have to illuminate our spirits', said Ailton Krenak at a September 1989 meeting in Washington. This leader of the Union of Indigenous Nations in Brazil spoke eloquently about how the planners of the planet were out of touch with the people. 'To just survive is very little', Krenak explained. 'To accept survival is to die a little. We want life. People everywhere are transforming their lives into a house, a car, objects. But people must choose between living and doing a "project".'

'We have to fall in love with the earth', is how Daniel Martin put it. This Roman Catholic missionary told *U.S. News & World Report* that 'when you fall in love, your ethical structure changes and you act differently, not because of an act of will, but because you see things in a new light'.

This kind of talk usually makes scientists uncomfortable. Cold, hard facts, after all, are the tools of their trade. Science should be as far from the church as the state is in most countries. But recently scientists and those pursuing spiritual matters, whether from the pulpits of established churches or in one-person ministries, have found themselves on the same path.

'The ecological crisis is a moral issue', said Pope John Paul II in his message for the World Day of Peace in 1990 (see Box 7–3). In this first papal message to be devoted entirely to environmental issues, the Pope's recitation of a litany of ecological woes made him sound more like Gro Harlem Brundtland than the leader of the Roman Catholic Church. (Unfortunately, his position on birth-control methods the church disapproves of is seriously at odds with his strong stand on the environment.)

Indeed, a few weeks later, Mrs Brundtland was talking about the spiritual challenge:

> A new environmental ethic must enter our consciousness, whether we are active in political or economic planning, whether we are bankers or industrialists, journalists or clergy, scientist or sales manager.
>
> The world's great spiritual and religious movements bear a profound responsibility because of the special influence, even authority, they hold in affecting people's most personal aspirations and motivation.
>
> Because of their strong, humane influence within vast communities in the world and because, by the nature of things, they cherish a benign role for Humanity in relation to the natural order, they have a great role to play in redirecting human motivation in relation to the future.

This speech was delivered in Moscow to the Global Forum on Environment and Development for Human Survival, a gathering where these two strands—a greener clergy and a more spiritual environmental movement—found common ground. A similar Global Survival Conference in April 1988 brought 200 spiritual and

Box 7–3.   Pope John Paul II on the Environment

In our day, there is a growing awareness that world peace is threatened not only by the arms race, regional conflicts and continued injustices among peoples and nations, but also by a lack of due respect for nature, by the plundering of natural resources and by a progressive decline in the quality of life. . . .

Faced with the widespread destruction of the environment, people everywhere are coming to understand that we cannot continue to use the goods of the earth as we have in the past. The public in general as well as political leaders are concerned about this problem, and experts from a wide range of disciplines are studying its causes. Moreover, a new ecological awareness is beginning to emerge which, rather than being downplayed, ought to be encouraged to develop into concrete programmes and initiatives. . . .

People are asking anxiously if it is still possible to remedy the damage which has been done. Clearly, an adequate solution cannot be found merely in a better management or a more rational use of the earth's resources, as important as these may be. Rather, we must go to the source of the problem and face in its entirety that profound moral crisis of which the destruction of the environment is only one troubling aspect. . . .

We cannot interfere in one area of the ecosystem without paying due attention both to the consequences of such interference in other areas and to the well-being of future generations.

The gradual depletion of the ozone layer and the related 'greenhouse effect' has now reached crisis proportions as a consequence of industrial growth, massive urban concentrations and vastly increased energy needs. Industrial waste, the burning of fossil fuels, unrestricted deforestation, the use of certain types of herbicides, coolants, and propellants: all of these are know to harm the atmosphere and environment. The resulting meteorological and atmospheric changes range from damage to health to the possible future submersion of low-lying lands.

parliamentary leaders to Oxford, England, including Mother Teresa, the Dalai Lama, Carl Sagan, and Soviet scientist Evguenij Velikhov.

By 1990 in Moscow, some 1,000 people from eighty-three countries gathered to hear almost every leading environmentalist, 'green' politician, and religious authority in the world. The week-long meeting listened to major addresses by Secretary-General de Cuellar, the Very Reverend James Parks Morton, US Senator Al Gore, Lester Brown, Victoria Chitepo, William Draper, Elie Wiesel, and—in a dramatic final session in the Kremlin—Mikhail Gorbachev.

While in some cases the damage already done may well be irreversible, in many other cases it can still be halted. It is necessary, however, that the entire human community—individuals, States and international bodies—take seriously the responsibility that is theirs. . . .

The concepts of an ordered universe and a common heritage both point to the necessity of a more internationally coordinated approach to the management of the earth's goods. . . . Recently there have been some promising steps towards such international action, yet the existing mechanisms and bodies are clearly not adequate for the development of a comprehensive plan of action. . . .

Modern society will find no solution to the ecological problem unless it takes a serious look at its life style. In many parts of the world, society is given to instant gratification and consumerism while remaining indifferent to the damage which these cause. As I have already stated, the seriousness of the ecological issue lays bare the depth of man's moral crisis. If an appreciation of the value of the human person and of human life is lacking, we will also lose interest in others and in the earth itself. Simplicity, moderation and discipline, as well as a spirit of sacrifice, must become a part of everyday life, lest all suffer the negative consequences of the careless habits of a few.

Source: Excerpted from 'Peace with God the Creator, Peace with All of Creation', Message of His Holiness Pope John Paul II for the Celebration of the World Day of Peace, 1 January 1990.

On the first day of the Global Forum, twenty-three well-known scientists, including three Nobel Prize winners, signed an appeal drafted by Carl Sagan for religion and science to join hands in preserving the earth. 'Many of us have had profound experiences of awe and reverence before the universe', they noted. 'Efforts to safeguard and cherish the environment need to be infused with a vision of the sacred.' One hundred religious leaders joined in the appeal, welcoming it as a unique opportunity to break down historic antagonisms between science and religion.

New links like those forged in Moscow among parliamentarians, scientists, and spiritual leaders are sure to translate into a greater effort to bring some spiritual guidance to the environmental movement and even more concern about ecology to the churches. The Declaration issued at the end of the Moscow meeting noted that 'we must find a new spiritual and ethical basis for human activities on Earth: Humankind must enter into a new communion with Nature, and regain respect for the wonders of the natural world'.

Jonathon Porritt, former head of UK Friends of the Earth, has captured well the potential impact of this new source of motivation for change. In a 1988 article he noted:

> We have reached the stage where the spiritual dimension of the ecology movement is as likely to be motivating people as its political dimension. . . . I think one can still discern today an enormous need for some kind of spiritual fulfilment and spiritual meaning, a need to look beyond some of the material confines within which most of our life is conducted. It's difficult to pin down that kind of feeling. . . .
>
> I do not believe the spiritual dimension of the Green movement is a recipe for taking oneself out of life and politics. Spirituality must not be seen as a retreat or a disengagement, but rather as an inspiration for political work. The spiritual dimension is the most compelling reason why we should be involved in politics. One cannot sustain that which one does not revere.

8

The Unfinished Agenda

'All this is to the good but it is not yet good enough.' This one sentence neatly summarizes the importance of the developments discussed in the preceding seven chapters. It was heard in Tokyo, in a September 1989 speech by Shridath Ramphal, Secretary-General of the Commonwealth and a former member of the Brundtland Commission. Mr Ramphal had just given a list of developments since the Commission finished its work, a list much like the contents of this book. 'There is some basis at least for hope', he concluded, 'if not for optimism'.

In these days of rapid political change, hope is abundant. In a January 1990 speech in Moscow, Gro Harlem Brundtland spoke of her own optimism: 'Never have we possessed greater collective capacity. Never has our knowledge of technologies and resources and how they interact been greater. And today, never have the political realities been more conducive to cooperation for sustainable development, across and between people.' Yet she also noted that 'we cannot address the real issues of today's world by focusing on simple terms such as East–West or North–South. We must, and we can, take a truly global perspective towards the year 2000 and beyond.'

This chapter draws on interviews with and letters from a number of people who take that global perspective, people who answered a request from the Centre for Our Common Future to comment on 'the unfinished agenda'. They are not all as hopeful as Mr Ramphal or Mrs Brundtland.

In an interview a few weeks before his appointment as Secretary-General of the 1992 Conference on Environment and Development, Maurice Strong seemed taken aback when asked for a list of the items on the unfinished agenda. 'Unfinished?' he said. 'It's not only unfinished, it's barely begun! What is needed is a fundamental change in our mindset, a radical revolution in the way we do things.' But then he added, sounding a bit more like his former colleagues on the Commission, 'we are not much beyond square one, but at least we are *at* square one'.

Where is square one? It is a place where more people are living in absolute poverty than a decade ago, with two thirds of them being under the age of 15. Where at least eleven million hectares of tropical forests disappear each year, and perhaps 35,000 species are wiped out, many undiscovered. Where one billion people—nearly 20 per cent of the world—are diseased, in poor health, or malnourished. Where an estimated twenty-four billion tonnes of topsoil are lost through erosion annually. Where the per capita income in the forty-two poorest countries is about $200, lower in many cases than it was at the start of the eighties. Where the burning of fossil fuels adds nearly six billion tons of carbon to the earth's atmosphere a year. Where nine million children die annually of illnesses that could be easily and inexpensively prevented—nearly 25,000 *children a day* dying needlessly. And where in 1990, close to ninety million people will be added to the 5.2 billion already trying to cope with this daunting picture.

We have a better understanding now of the reasons we need to move towards sustainable development if we are to get beyond square one. But we have not figured out the best ways to do so. Commenting on the unfinished agenda, Mostafa Tolba, Executive Director of the UN Environment Programme (UNEP), noted: 'There is no doubt the World Commission on Environment and Development helped define *why* environmentally sound and sustainable development is essential for our planetary survival. We must now define *how*—how to build truly global co-operation, how to finance public policy measures on an international scale, how to translate the words "sustainable development" into action.'

As mentioned throughout this book, the agenda item on which the least progress has been made is the elimination of poverty; the eighties have been rightly called a lost decade in terms of development. Yet, few of the new converts to environmentalism, those proclaiming that the nineties will be the 'green' decade, see the connection between their concern about the health of the earth and the need to eliminate poverty. As the Brundtland Commission pointed out: 'Poverty is a major cause and effect of global environmental problems. It is therefore futile to attempt to deal with environmental problems without a broader perspective that encompasses the factors underlying world poverty and international inequality.'

Janet Hunt of the 'One World or . . . None' campaign in Australia is one person well aware of this link. In her letter about the signs of hope, or lack of them, she wrote: 'Governments like Australia's are prepared—even forced by public opinion—to respond to our domestic environmental crisis. But who is prepared to increase aid and do something about the trade injustices which perpetuate such inequality which is at the heart of so much of this overexploitation? Not many. There is no international movement to really tackle the poverty question at the governmental level.'

How can we launch such an international movement? How can we get policy-makers to see the futility of attempts to 'save the earth' that do not at the same time end the poverty of those who live on it? 'The poor', noted Pope John Paul II in his 1990 message on the environment, 'to whom the earth is entrusted no less than to others, must be enabled to find a way out of their poverty. . . . the proper ecological balance will not be found without *directly addressing the structural forms of poverty* that exist throughout the world.'

Ever since Indira Gandhi's moving speech at the Stockholm conference in 1972, poverty has been called the greatest pollutant. As the Brundtland Commission explained: 'Those who are poor and hungry will often destroy their immediate environment in order to survive: They will cut down forests; their livestock will overgraze grasslands; they will overuse marginal land; and in growing numbers they will crowd into congested cities. The cumulative effect of these changes is so far-reaching as to make poverty itself a major global scourge.'

Maybe we should apply to poverty an important lesson industry has learned over the last decade: it costs less in the long run to avoid pollution from the start than to clean up, one by one, the environmental problems it creates. One of the first to discover this was the 3M Company, which claims that since 1975 it has saved well over $1 billion in its Pollution Prevention Pays programme.

By the same token, it pays to invest in literacy campaigns, especially for women, so that people have a better chance to participate fully in the economic life of their societies. It pays to provide clean water and primary health care, so that people can work without the constant drain of disease. It pays to invest in low-cost, appropriate technologies, so that people can produce the goods they need in rural areas, where poverty is pervasive. It pays to eliminate agricultural subsidies and trade barriers in industrial nations, so that Third World farmers can earn a fair price for their goods. It pays to invest in making family planning widely available, so that people who wish to have fewer children are able to do so. It pays to give people legal rights to their land and their homes, so that they know the efforts they make to conserve or improve them will benefit themselves and their children.

It pays to invest in land-reform programmes . . . in innovative schemes to extend credit to rural communities . . . in small-scale enterprises in urban and rural areas . . . in low-cost energy-saving stoves . . . in all the components of a poverty-alleviation strategy. It pays, in other words, to give the poor better access to land, food, health care, education. Otherwise, the contribution that one fifth of humanity could make to our common future is virtually lost. As 3M found out, it costs a little more in the beginning to avoid pollution. But in the long run, pollution prevention pays.

An important component of a strategy to set individual nations and the global economy on a more sustainable path is a workable resolution of the debt crisis. Some hopeful steps have been taken by governments that have erased from the books the bills of the poorest nations. But that has scarcely touched the surface of the Third World's outstanding debt—over half of it to private banks—of $1,165 billion at the end of 1989. For the last six years, developing countries have been paying more to their creditors each year than they receive in new aid. In 1988, this net outflow topped $50 billion—a situation roundly deplored as 'perverse', but one that remains unresolved none the less.

In a 1988 speech in Mexico City, Mrs Brundtland delivered a blunt assessment of the debt crisis:

> Let us be frank about this. It will not make sense to demand that all the debt be repaid. To maintain such an iron code will hit the most vulnerable social groups in many debtor countries. It will increase the pressure on the environment too, as natural resources are cashed in to service debt. It will act as a serious obstacle to investments and innovations and it will prolong the time until the debtor countries can assume their rightful position in the international economy. The

ultimate goal must be to forge an economic partnership based on equitable trade.

Since then, plans to ease this problem have come and gone. The World Bank reports that only $14 billion in debt relief was arranged in 1989, compared with $22 billion in 1988. A comprehensive plan for debt relief, for rescheduling of payments, and for conversion of some loans to grants should be included in any discussions of a global action plan on the environment. The need to do so may become even clearer as plans unfold for the UN conference in Brazil, the largest debtor in the Third World, where the inflation rate in 1989 was an unfathomable 1,765 per cent. Under President Sarney, Brazil stopped paying interest on its debt in July 1989.

Beyond the need to relieve the debt burden, additional funds must be devoted to moving global economic activity on to a sustainable track. The pressing financial needs of the next decade were raised by nearly all those contacted for comments on the unfinished agenda. Mr Ramphal, for example, noted that 'it is becoming clear that a major commitment needs to be made by the developed world in terms of resource flows to support development and more specifically to assist conservation measures in relation to the ozone layer, climate change, tropical forests, protection of species, etc. The scale of this funding and the mechanisms are emerging as a major sticking point in negotiations.'

The longer these negotiations take, the more it will cost to repair the damage and to change the direction of development. As discussed in Chapter 7, several innovative funding mechanisms are being considered—carbon taxes, fees for use of the global commons, a climate fund, a planet protection fund. And the recently much-touted peace dividend may provide some support. Soviet Foreign Minister Shevardnadze has noted that 'the establishment of an international ecological fund, formed by deductions from resources released by cuts in military spending in the course of disarmament, is a practical and feasible possibility'.

Another issue raised in Chapter 7—appropriate technology—is also high on many people's list of unfinished business, in terms of the need for environmentally sound technology in developing countries. As Dr Tolba put it, 'Defining and promoting clean technologies is one side of the coin. The other side is getting them where they are needed. Of critical importance is technology transfer, so that environmentally benign technologies can be widely used in developing countries.'

In some cases, the Third World has an advantage over the First when it comes to sustainable industrial development. If support is forthcoming in the form of funding and open access to the most energy-efficient, materials-conserving technologies, industry can start off on the right foot. Refrigerators can be manufactured without using chlorofluorocarbons in the insulation or the coolant, to cite one small example. But in the industrial countries, existing operations must be modified, or perhaps scrapped. Ample precedent exists to show that change can occur—witness the record-breaking improvement in the energy efficiency of the Japanese economy since 1973—but it will be more in the nature of an adjustment, and less a chance to do things right from the start.

For industrial countries, structural adjustments also lie ahead in agriculture, as farmers continue to be paid more for their produce than they would be in an open world market. The agricultural sector in western nations is highly subsidized, to the tune of $100 billion a year, and highly protected. One result with environmental repercussions has been the over-use of pesticides, as farmers rush to squeeze as much out of a given land area as possible. In a 1989 speech in Tokyo, the difficulty of taking bold steps to remedy this was explained by Mr Ramphal—not, it should be noted, someone in office at the moment who has to answer to voters:

> A change of direction will require governments to stand firm against powerful pressure groups: the farmers of the North; the urban masses of the South. The implication may well be that in the transition to a saner world agriculture, food prices should rise. Governments will have the awesome task of persuading their people that, in this and other ways, sustainable development offers no easy options and may be very painful. In fact, at present, many developing country governments are being forced—through structural adjustment conditionality—to inflict this pain; but in the developed world, where per capita consumption of energy, food and other resources is very much higher, 'green' rhetoric is not being matched by real action.

This hints at some of the changes that lie ahead in life-styles, a political hot potato that few politicians are willing to touch. The 'pain' of sustainable development—for farmers and city-dwellers, for rich nations and poor—could be eased by political leaders with vision, leaders who see beyond elections tomorrow or the day after to the legacy they would leave for their children by helping us face up to hard choices. A few of these individuals are now being heard, and no doubt this year a few more will chime in. But a few is not enough. What is missing is the political consensus that these issues demand the world's undivided attention if we are to leave our children an earth that will support them.

*   *   *

These general points have been about the policy imperatives for the nineties: poverty alleviation, resolution of the debt crisis, the Third World's need for additional funds and for efficient and appropriate technology, an honest look at the path towards a sustainable society. Another item on many unfinished agendas is more specific—applying the Brundtland Commission's recommendations to regional and national situations. A complaint often heard after the report's release was that *Our Common Future* lacked a regional perspective, that it painted a picture of the changes needed with too broad a brush. Indeed, the prerequisites the Commission laid out for the pursuit of sustainable development were intentionally broad:

- a political system that secures effective citizen participation in decision making,
- an economic system that is able to generate surpluses and technical knowledge on a self-reliant and sustained basis,
- a social system that provides for solutions for the tensions arising from disharmonious development,
- a production system that respects the obligation to preserve the ecological base for development,
- a technological system that can search continuously for new solutions,
- an international system that fosters sustainable patterns of trade and finance, and
- an administrative system that is flexible and has the capacity for self-correction.

These still hold true, and, as mentioned in Chapter 1, progress has been seen on the first prerequisite throughout Eastern Europe, and has been hinted at in South Africa.

One remedy for the broad brush-strokes of the Commission has been regional meetings about *Our Common Future*. When the United Nations welcomed the report in December 1987, it invited governments to join its regional commissions and the UN Environment Programme in national and regional follow-up conferences. The first of these, in Uganda, was held in June 1989. Thirty-five countries were represented by Ministers of Environment, of Planning, and of Education, and by representatives of youth, women's, and other non-governmental organizations. In a strongly worded Kampala Declaration, they agreed that 'development which is not sustainable should no longer be called development', and adopted an agenda for action towards sustainable development in Africa (see Box 8-1).

The next region to consider the implications of the report was the Economic Commission for Europe. (By a quirk of UN logic, this includes Canada and the United States.) Ministers and their equivalents from some thirty-four countries met in Bergen, Norway, in May 1990 for a conference on 'Action for a Common Future'. Their deliberations focused on four overall topics that had been discussed in preliminary workshops: the economics of sustainability, awareness-raising and public participation, sustainable energy use, and sustainable industrial activity.

Similar meetings are planned for later in 1990 in the other two regions within the UN system. The Economic Commission for Latin America and the Caribbean plans to hold an August conference in Santiago, and a Ministerial Conference on the Environment in Asia and the Pacific is to be held in Bangkok in October.

Adding to these important efforts to develop regional perspectives on our common future is a recent Latin American initiative outside the UN system, though it was established with the help of several branches of the United Nations. Fourteen prominent politicians and scientists announced in October 1989 the formation of a new Commission on Development and the Environment (see Box 8-2). It is led by Oscar Arias Sanchez of Costa Rica, UN Development Programme Assistant Administrator Augusto Ramirez Ocampo, and Inter-American Development Bank President Enrique Iglesias. Included on the Commission are the former Presidents of Colombia, Ecuador, and Mexico, and three members of the Brundtland Commission.

The Commission plans to publish a report by June 1990 on priorities and strategies for Latin America and the Caribbean. As the lack of this perspective in *Our Common Future* has been a particular source of criticism, the new report is especially welcome.

Box 8–1. Sustainable Development in Africa

■ We undertake to integrate environmental concerns into all existing and future economic and sectoral policies to ensure that they protect and improve the environment and natural resource base on which the health and welfare of our people depend. We must also begin to implement new sustainable development programmes that increase our possibilities for meeting the pressing needs of our people today without compromising the prospects of future generations. . . .

■ We further commit ourselves to developing African strategies and technologies for production, preservation, storage, distribution and consumption which will stimulate sustainable economic growth and secure livelihoods in the rural areas where the majority of our populations live. We should at the same time adopt common strategies concerning imported technologies which could adversely affect our environment.

■ In the context of reviving economic growth with greater equity and meeting the essential needs for food, water, energy and jobs of our people, we are committed to immediate action on the following priority issues and goals for achieving sustainable development in our countries and continent.

- Managing demographic change and pressures
- Achieving food self-sufficiency and food security
- Ensuring efficient and equitable use of water resources
- Securing greater energy self-sufficiency
- Optimizing industrial production
- Maintaining species and ecosystems
- Preventing and reversing desertification

■ To move from the present and often destructive processes of development towards sustainable development will require a transition period of years to decades. The duration and success of that transition will depend on a strong and continuous political commitment at the highest levels within and among our countries, on the active role of an informed, involved public and on pragmatic programmes of national action and subregional and regional co-operation.

■ We have therefore agreed on 'Priorities for Immediate Action' for the seven priority issues and goals for moving towards sustainable development in Africa. We commit ourselves from today to begin to implement them immediately within and among our own countries.

■ We call upon the international community to support our efforts in the spirit of true partnership among States in providing for our common future.

Source: Excerpted from 'The Kampala Declaration on Sustainable Development in Africa', adopted by the First African Regional Conference on Environment and Sustainable Development, Kampala, Uganda, 12–16 June 1989.

Box 8–2. Development and the Environment in Latin America and the Caribbean

■ Humanity desires to consolidate peace. We are pleased with the relaxation of tensions between the superpowers and with the favorable tone that this brings to international co-operation. We welcome the resurgence of multilateral co-operation and the strengthening of the central role of the United Nations. These developments create propitious circumstances for concerted action and attention to two major, pressing global issues: development and the environment.

■ The World Commission on Environment and Development represented a positive step in this direction, in calling for the construction of a new era based on sustainable development and eradication of poverty. The exercise that we have begun responds to the recommendations of that Commission.

■ In Latin America and the Caribbean, the people and governments are conscious of the need to launch a new development strategy. Various positive actions have begun in our region, promoted by its political leadership or by organizations and people who come from Latin American society, acting within the framework of agreement among nations. . . .

■ The present economic crisis and the environmental threat are rooted in defective models of development—the economy of opulence and waste of the North and the economy of poverty and survival of the South. Thus, the challenge now is to design a strategy of development that generates welfare for all the population, and in harmony with nature: Development integrated with justice and conservation. . . .

The Commission has no intention of disbanding after this initial report, however, and will continue to contribute a much-needed regional view of sustainable development.

All these developments since the publication of *Our Common Future* and of UNEP's *Environmental Perspectives to the Year 2000* are occurring at a time when the United Nations seems to have received a new lease on life. One indication of the revitalization is the US government's February 1990 announcement that it finally wanted to pay its dues. President Bush's budget proposal included a plan to wipe out its debts to the United Nations and its specialized agencies over the next five years, although it is unclear if this goodwill gesture can survive congressional negotiations on the

■ Regional integration will permit us to make the most of our comparative advantages in human, cultural, natural and genetic resources. We must revise our own agenda of development and environment within a regional perspective. Latin America and the Caribbean can enrich the global debate over development and environment, within a framework of principles of sovereignty, universality, and solidarity, as we face questions that affect the entire human race. . . .

■ In the area of co-operation on global ecological matters, the following will be crucial: a major financial contribution from the industrialized countries, and rapid and growing technology transfer. . . .

■ We support the call for an early North–South summit and ask that a priority concern be a common strategy on development and the environment. The strengthening of co-operation between developing and developed countries is crucial to achieve international agreements on the urgent global environmental risks.

■ We consider that technical and political reflection on the links between opulence, poverty, population, environment and natural resources will better permit us to define proposals for our region. We should mobilize public opinion around this task and draw in political leadership and the private sector. This joint enterprise should form the basic accord that guides the future development of our hemisphere and should not be the monopoly of either a political party or an ideology, nor of one nation. Thus we can contribute to creating a better and more hopeful destiny for the population of Latin America and the Caribbean.

Source: Excerpted from First Communiqué, Commission of Latin America and the Caribbean on Development and the Environment, New York, 3 October 1989.

overall budget. Until this debt is cleared, the United States has the largest outstanding bill to the main UN budget, followed by South Africa and Iran.

This support, if it materializes, comes at an important time in international relations. As Shridath Ramphal pointed out in his speech in Tokyo, 'after a long, frustrating, period in which multilateralism was in retreat, environment has provided the trigger for a major burst of activity, much of it led by UNEP, in formulating international principles and rules'. This activity will reach a major turning-point two years from now in Brazil, at the United Nations Conference on Environment and Development, set on the twentieth anniversary of the UN Conference on the Human Environment, the Stockholm Conference.

In its resolution calling for the conference to be held, the

UN General Assembly stressed that poverty and environmental degradation are closely interrelated and that environmental protection in developing countries cannot be considered in isolation from the development process. Twenty-three objectives for the meeting were specified, including promoting environmental education and examining the possibility of a special international fund to aid in the transfer of environmentally sound technologies to the Third World.

The General Assembly called for the highest possible level of government participation in 1992, a call echoed a month later in Moscow by President Gorbachev. Speaking at the Global Forum on Environment and Development for Human Survival, he indicated that the Soviet Union wanted the Brazil Conference to be a summit meeting. He added: 'It would probably be right if the issue of drawing up an international code of ecological ethics were raised there. . . . Such an undertaking would symbolize a readiness on the part of the world community, in the person of its top representatives, to construct life in the 21st century according to new laws. The 1992 conference could also adopt a global action programme on environmental protection and rational use of natural resources.'

These goals are admirable. Whether they are also realistic is not yet clear. The Brazil Conference will be a critical chance—perhaps the world's last—to change the course of development. It is loaded with opportunities, and with dangers.

The danger is that, when push comes to shove, the industrial nations of the world will fail to match the fine rhetoric of their leaders about bearing responsibility for the poor state of the earth, and therefore responsibility for getting the world on a path towards sustainable development. The new global environment is one of co-operation, but does the industrial world mean what it says about paying its dues? The mandate is clear. Dr Tolba told the UN General Assembly: 'Let us be in no doubt that this cannot be a conference like any other, one at which more is *said* than *done*. The time to be satisfied with a little, because we believed we could not accomplish everything, is over.'

In the interview he gave before he was asked to become the Secretary-General of the meeting, Maurice Strong also spoke about the dangers: 'There is a real possibility of a breakdown between North and South at the 1992 conference, however, if the North is not willing to come through with real financial resources. This could set the world back ten to twenty years from making the changes that are needed—or could even make them impossible.'

Mr Ramphal voiced similar concerns in September 1989, concerns that have become more pressing for many in the Third World in light of subsequent developments: 'For Western industrial countries, this is a crucial time that could easily be wasted basking in a warm glow of smug complacency because the rest of the world now wishes to follow their political and economic example. If they spurn the cries for help from the developing world—and from Eastern Europe—if they treat the issues of environment and development in a parochial, self-serving manner, they could reap a bitter harvest.'

But money alone is not the answer. It is necessary, but it is not enough. Hand-outs help people survive, but they do not always let them prosper. In a February 1990 interview about his hopes for the UN Conference, Fabio Feldman—one of Brazil's leading environmentalists, and now an elected Federal Deputy in the Government—noted:

> What must happen in 1992 is a change in the relationship between North and South. Not just a flow of money, but a real change in the *nature* of the relationship. There must be a change in trading patterns, in which goods from the Third World have always been devalued. And the North must guarantee access to its technologies. If the result of 1992 is just more money, in another ten years the world will be back to the difficult situation it is in today.

His concerns are matched by those of UNEP Executive Director Tolba. The result of discussions in Brazil on new financing mechanisms, he wrote, should be 'a truly global partnership, in which all countries contribute, in which all countries have an equal say in resource allocation. When such a global partnership takes shape, then the "unfinished agenda" will begin.'

*   *   *

Tokyo, February 1987. The World Commission on Environment and Development gathers around another hotel conference table. This last meeting of the Commission has been the longest, filled with tension as people who have become friends over nearly three years of discussions struggle to reach unanimous agreement on the final text. But agree they do. And they are quite certain that in addition to sounding an alarm about our future being threatened by current patterns of development, they want the world to know that progress has been made . . . that they found grounds for hope . . . that 'people can cooperate to build a future that is more prosperous, more just, and more secure'.

Three years later, their hope does not seem misplaced. The progress towards our common future documented in these pages can be called a start. A move in the right direction. Indeed, all this is to the good. But it is not yet good enough.

Appendix 1

The Centre for Our Common Future

The final Report of the World Commission on Environment and Development, *Our Common Future*, met with universal praise when it was published in April 1987, and was soon seen as an invaluable tool in the global effort to achieve sustainable development. In the belief that the opportunity to bring about real change could be lost if the momentum created by the Report was not maintained, strengthened, serviced, and encouraged, the Centre for Our Common Future was launched in April 1988. It quickly became a focal point for groups and individuals seeking to use the Report in their work.

While its principal purpose is to activate and extend the global debate on sustainable development, the Centre also acts as an independent clearing-house for the exchange of ideas and activities, providing a constant flow of information, advice, and encouragement. It plays a useful and effective role in uniting the various sectors of human activity towards common goals.

Working on a small budget, with funding provided by voluntary contributions, support for its creation and continued operations has come from governments, private foundations, and its Working Partners. The International Union for Conservation of Nature and Natural Resources and the International Institute for Environment and Development provided valuable support and assistance with its creation.

The need for systematic reporting of the many responses to the Report and initiatives based on it soon became clear. The first issue of the Centre's quarterly publication, the *Brundtland Bulletin*, was published in September 1988, and it quickly grew from forty-four pages in length to 120 pages by the sixth issue, in December 1989.

The small staff at the Centre, which is located in the offices used for three years by the World Commission, has responded to thousands of requests from around the world for printed and video materials in the two years of its existence.

To build a broad base of support for the implementation of the Report's recommendations, key environment, development, media, trade union, youth, women's, industry, and financial organizations were asked to join the Centre as Working Partners. Included are intergovernmental and non-governmental organizations (NGOs), foundations, academic groups and research centres, and professional associations. They range from NGOs in the Third World such as the Environment Liaison Centre International (Nairobi), Asian NGO Coalition (Manila), and Fundacion Ambiente y Recursos Naturales (Buenos Aires) to distinguished bodies such as the USSR Academy of Sciences (Moscow), World Resources Institute (Washington, DC), and the Third World Academy of Sciences (Trieste, Italy). Many have been designated as repositories of archive materials from the Commission along with current materials from the Centre itself. By March 1990 the Centre had nearly 150 Working Partners around the world.

The enormous continuing interest in *Our Common Future* led the Centre to commission this book to summarize the initiatives to date. It draws on the extensive files kept in Geneva, and readers wishing further information on any initiatives mentioned are encouraged to contact the Centre for references and follow-up details.

One important event in which the Centre will play a key role is the UN-sponsored Conference on Environment and Development, to be held in Brazil 5–19 June 1992. The General Assembly resolution on the Conference calls for the highest possible level of government participation. It sets out a broad range of objectives for the meeting, noting that environmental issues will need to be addressed within the development context. In inviting all States to take an active part in the preparations for the Conference, the resolution specifically asks them to promote broad-based national preparatory processes involving the scientific community, industry, trade unions, and concerned non-governmental organizations. NGOs in consultative status with the Economic and Social Council were also asked to contribute to the meeting.

To mobilize its Working Partners and others to contribute both to the preparatory process and to the Conference itself, in March

1990 the Centre hosted the first meeting of the constituencies outside government to discuss the 1992 Conference. Some 100 individuals from forty countries attended this meeting in Vancouver.

*   *   *

The World Commission on Environment and Development was created as an independent body in 1983 by the United Nations. The Commission's mandate during its three years of existence was to re-examine the critical issues of environment and development and formulate new and concrete action proposals to deal with them, to assess and propose new forms of international co-operation that could break out of existing patterns and foster needed change, and to raise the level of understanding and commitment everywhere.

From the very beginning, the Commission agreed that its processes would be open, visible, and participatory. Its Public Hearings, held in fourteen cities around the world, became its unique feature, its 'trademark', demonstrating that the issues they addressed are indeed of global concern, transcending national boundaries and different cultures. Hundreds of organizations and individuals gave testimony during these Public Hearings and over 500 written submissions, constituting more than 10,000 pages of material, were received by the Commission in connection with them. These background materials have been compiled into an Archive Collection on Sustainable Development that is available at the Centre.

The members of the Brundtland Commission served in their individual capacities. The brief biographical details given here begin with their positions as of January 1990, and include the positions they held during the Commission or at the time *Our Common Future* was published.

Gro Harlem Brundtland. Chairman—Leader of the Norwegian Labour Party; former Prime Minister, Norway.

Mansour Khalid. Vice-Chairman—former Minister of Foreign Affairs, Sudan.

Susanna Agnelli. Under-Secretary of State for Foreign Affairs, Italy.

Saleh Abdulrahman Al-Athel. President of King Abdulaziz City for Science and Technology, Saudi Arabia.

Pablo Casanova.* Professor of Political and Social Sciences, National Autonomous University of Mexico, Mexico.

Bernard T. Chidzero. Minister of Finance, Economic Planning and Development, Zimbabwe.

Lamine Fadika. Minister of Marine Affairs, Côte d'Ivoire.

Volker Hauff. Mayor of Frankfurt, Federal Republic of Germany; former Member of Parliament, Federal Republic of Germany.

Istvan Lang. Secretary-General of the Hungarian Academy of Science, Hungary.

Ma Shijun. Director of Research Centre of Ecology, Academia Sinica, People's Republic of China.

Margarita Marino de Botero. President, El Colegio de Villa de Leyva, Colombia; former Director General of the National Institute of Renewable Natural Resources and the Environment, Colombia.

Nagendra Singh. Former President of the International Court of Justice, The Hague, Netherlands (deceased, December 1988).

Paulo Nogueira-Neto. Professor of Ecology, University of São Paulo, Brazil; former Federal Secretary of the Environment, Brazil.

Saburo Okita. Chairman, Institute for Domestic and International Policy Studies, Japan; former Foreign Minister, Japan.

Shridath S. Ramphal. Secretary-General of the Commonwealth of Nations, United Kingdom.

William Doyle Ruckelshaus. Chief Executive Officer, Browning-Ferris Industries, United States; former Administrator of the Environmental Protection Agency, United States.

Mohamed Sahnoun. Algerian Ambassador to Morocco; former Algerian Ambassador to the United States.

Emil Salim. Minister of State for Population and the Environment, Indonesia.

Bukar Shaib. Former Minister of Agriculture, Nigeria.

---

* In August 1986, for personal reasons, Pablo Casanova ceased to participate in the work of the Commission.

Vladimir Sokolov. Member USSR Academy of Science, Union of Soviet Socialist Republics.

Janez Stanovnik. President, Socialist Republic of Slovenia, Yugoslavia; former member, Presidium of the Socialist Republic of Slovenia, Yugoslavia.

Maurice Strong. Secretary-General, United Nations 1992 Conference on Environment and Development; Under-Secretary General of the United Nations.

Jim MacNeill (*ex officio*).—Director, Sustainable Development Programme, Institute for Research on Public Policy, Canada; former Secretary General of the World Commission on Environment and Development.

* * *

The Centre for Our Common Future
Palais Wilson
52, rue des Paquis
CH-1201 Geneva, Switzerland
Tel. (022) 732 7117 Fax (022) 738 5046

Appendix 2

Glossary of Environmental Issues and Terminology

Acid Rain, Acid Precipitation. Popular term for rain, snow, sleet, clouds, and fog that contain an unnatural amount of sulphur dioxide, nitrogen oxides, and other pollutants due to the burning of fossil fuels in power plants, factories, and motor vehicles. Although factories and heating-plants in industrial countries added tall smoke-stacks over the last few decades to disperse these emissions, the pollutants have only landed further afield, and across national boundaries. Acid rain has damaged trees, lakes, and soils throughout Europe and parts of North America.

Basle Treaty. Approved on 22 March 1989 in Basle, Switzerland, this accord governs the international transport and disposal of hazardous waste. An underlying principle is that wastes and the hazards associated with them must be minimized. Signatories agree, among other things, to obtain written permission from an importing country before a shipment of such materials leaves the exporting country, which in turn must be assured that disposal will occur in an environmentally sound manner. Thirty-four countries and the European Economic Community signed the accord right away; the treaty is expected to enter into force before the end of 1990, after twenty countries ratify it.

Bergen Conference. Regional follow-up conference to the Brundtland Report organized by the Norwegian government in co-operation with the Economic Commission for Europe, a UN group that includes Canada and the United States. Under the theme 'Action for a Common Future', the ministerial-level meeting was held in Bergen, Norway, 8–16 May 1990. From 5 to 7 May, a 'Youth Action for a Common Future' conference was held, followed on 8–12 May by a meeting of experts on 'Sustainable Development, Science and Policy'. Bergen was also the site of an international environmental-technology exhibition called 'Worldcare Action 1990', from 8 to 11 May.

Biodiversity. A term for the variety of ecosystems, plant and animal species, and genetic differences that exist on earth. Scientists estimate the number of species as between 5 million and 30 million—a range that indicates how poor our knowledge of life on earth is.

Brazil Conference. A meeting scheduled for 5–19 June 1992, on the twentieth anniversary of the Stockholm Conference. The official title of the meeting is the 'United Nations Conference on Environment and Development'. Maurice Strong, a Canadian who led the Stockholm meeting in 1972 and who was UNEP's first Executive Director, was appointed as Secretary-General of the conference in February 1990; the secretariat is based in Geneva, with satellite offices in Nairobi and New York.

Chlorofluorocarbons (CFCs). A class of chemicals implicated in two major environmental problems: ozone-layer depletion and global warming. It takes about fifteen years for a CFC molecule to move up to the stratosphere, where it can last about 100 years and destroy over that time 100,000 ozone molecules. CFCs are used as coolants in refrigerators and air-conditioners, propellants in aerosol cans, solvents used during the manufacture of computers, and blowing-agents that inflate flexible foams. Their use as a propellant in aerosols was banned in Canada, Norway, Sweden, and the United States in 1978; other industrial countries now have partial or voluntary bans on such uses. All CFC uses are covered by the Montreal Protocol, and are likely to be halted in industrial countries by 2000 at the latest.

Climate Change. See *global warming.*

Debt-for-Nature Swaps. An arrangement in which a private group in an industrial country (so far, conservation groups in the United States) buys a certain amount of a developing-country government's debt at a discounted rate from a private creditor who has decided that the full amount is unlikely ever to be paid. In return, the government sets aside a specified tract of land as a nature reserve or other agreed-upon category of protected land, or establishes some conservation programme that it and the new holder of the debt have agreed on.

Deforestation. At least eleven million hectares of tropical forest are lost every year. In the tropics, ten hectares are cleared for every one planted. Although the causes vary by region, one estimate indicates that slash-and-burn agriculture and scavenging for wood-fuel, often in the wake of road-building for commercial purposes, account world-wide for 40–50 per cent of deforestation; grazing for 10–20 per cent; commercial agriculture for 10–20 per cent; the building of dams, mines, roads, and buildings for 10–15 per cent; forestry and plantations for 5–10 per cent; and forest fires for 1–15 per cent.

The adverse effects of deforestation include the loss of habitat for countless plant and animal species, many as yet undiscovered by scientists; the destruction of the homes and livelihoods of native tribes in many parts of the tropics; the denuding of mountainsides, providing an easy path for soils to wash away and rain to flood valleys below; and the addition of carbon dioxide to the atmosphere when the trees are cleared or burned, adding to the problem of global warming.

Global Warming. The anticipated increase of global average temperatures by 2.5 to 5.5 degrees Celsius (4.5 to 9.9 degrees Fahrenheit) in the next century. The greatest warming is expected near the poles, with little change anticipated near the equator. Among the more worrying consequences will be rising sea levels, which could have great impact on such developing countries as Bangladesh and Egypt; shifts in the needs and location of primary agricultural areas; and the acceleration of species loss, as plants and animals are unable to adjust quickly to changes in the conditions in their habitats.

Global warming is caused by the increased concentration of carbon dioxide and other gases in the atmosphere, which prevents some of the sun's heat from escaping and, hence, turns the world into one big greenhouse. About 45 per cent of this is due to the burning of fossil fuels. A further 10 per cent is caused by the clearing and burning of forests, which contain huge amounts of carbon. And some 45 per cent stems from the accumulation of methane, nitrous oxide, and CFCs in the atmosphere. Considering all these sources, the ten countries contributing most to global warming, in order of contribution, are the United States, the Soviet Union, China, Brazil, Japan, Indonesia, West Germany, the United Kingdom, India, and Colombia.

Hague Declaration. A statement on protection of the global atmosphere signed by seventeen heads of government and representatives of seven other governments following a meeting in The Hague on 10–11 March 1989. By early 1990 forty governments had agreed to the declaration. Among other items, it called for a new or newly strengthened institution within the United Nations to combat further global warming.

Halons. Used in fire-extinguishers, these chemicals contain the element bromine, an even more effective destroyer of the ozone layer than chlorine.

Intergovernmental Panel on Climate Change (IPCC). A committee of government representatives formed by the UN Environment Programme and the World Meteorological Organization in 1987 to tackle the issue of global warming. Working groups are preparing reports on three areas of concern: scientific information on all factors affecting climate change, the environmental and socioeconomic impacts, and policy issues raised by global warming. IPCC negotiations are expected to lead to a draft treaty on climate change by 1992.

Kampala Conference. The first regional follow-up conference to the Brundtland Report, organized by the Economic Commission for Africa and the UN Environment Programme. From 12 to 14 June 1989 an expert group met in Kampala, Uganda, and on 15 and 16 June a ministerial-level session was held. The conference adopted an Agenda for Action on Sustainable Development as well as issuing a Kampala Declaration at the conclusion of the meeting.

Montreal Protocol on Substances That Deplete the Ozone Layer. Signed by twenty-four nations and the European Community in September 1987, the agreement calls for CFC use to be reduced by 20 per cent by 1994 and by 50 per cent by 1999 (using 1986 levels as the base), and for halon use to halt within three years of the protocol coming into force, which it did on 1 January 1989.

Amendments to the protocol, in order to halt CFC use entirely, are quite likely and are now being negotiated.

Non-governmental Organizations (NGOs). A catch-all term that covers what are known in various nations as community, citizen, public-interest, or voluntary groups. Although trade associations (those representing the interests of specific industrial or professional sectors) are by definition non-governmental, the term NGO is not generally applied to them.

Ozone-layer Depletion. Ozone is the only gas in the upper atmosphere that limits the amount of harmful ultraviolet (UV) radiation reaching the earth. Since 1969 it has dropped 3.0–5.5 per cent during the winter at northern mid-latitudes. (Ozone in the lower atmosphere, on the other hand, is increasing and is a pollutant.) Increased UV radiation can suppress the immune system, leading to more severe infectious diseases, and can increase the number of non-melanoma skin-cancer cases and of cataracts. Higher doses of UV radiation also appear to reduce the yield of soybeans and to threaten phytoplankton and other minute organisms important in the oceans' food chain.

Stockholm Conference. The UN Conference on the Human Environment, held in Stockholm in June 1972. Seen by many as the start of the global environmental movement, the conference led among other things to the creation of the UN Environment Programme.

UN Development Programme (UNDP). This major UN organization was established in 1951. It currently supports 5,300 projects valued at $7.5 billion in 152 developing countries. UNDP resident representatives live in 112 of these nations. Projects supported span the full range of development needs—agriculture, education, employment, fisheries, health, industry, science and technology, transport, and so on. Eighty per cent of the programme funds go to countries with per capita incomes of $750 a year or lower. The headquarters is in New York; the Administrator is William H. Draper III, an American.

UN Environment Programme (UNEP). UNEP began operating in 1973, following its establishment by the General Assembly. Its small budget of $40 million in 1989 was prcvided through a voluntary Environment Fund to which half the members of the

United Nations contribute. UNEP is not an operational agency, but describes its role as catalytic and co-ordinating; it sometimes also acts as a political broker among governments, as it did in the negotiations regarding the ozone layer, helping forge agreements. The headquarters of UNEP is in Nairobi; the Executive Director is Dr Mostafa K. Tolba, an Egyptian.

Index